江西财经大学东亿学术论丛·第一辑

基于Copula的相关性测度

单青松 著

Measures of Association and
Dependence Through Copulas

经济管理出版社
ECONOMY & MANAGEMENT PUBLISHING HOUSE

图书在版编目（CIP）数据

基于 Copula 的相关性测度 / 单青松著. —北京：经济管理出版社，2019.10
ISBN 978-7-5096-6187-1

Ⅰ.①基⋯　Ⅱ.①单⋯　Ⅲ.①时间序列分析—研究　Ⅳ.①O211.61

中国版本图书馆 CIP 数据核字（2019）第 242531 号

组稿编辑：王光艳
责任编辑：李红贤
责任印制：黄章平
责任校对：董杉珊

出版发行：经济管理出版社
　　　　　（北京市海淀区北蜂窝 8 号中雅大厦 A 座 11 层　100038）
网　　　址：www.E-mp.com.cn
电　　　话：(010) 51915602
印　　　刷：北京晨旭印刷厂
经　　　销：新华书店
开　　　本：720mm×1000mm /16
印　　　张：8.5
字　　　数：148 千字
版　　　次：2019 年 12 月第 1 版　2019 年 12 月第 1 次印刷
书　　　号：ISBN 978-7-5096-6187-1
定　　　价：68.00 元

·版权所有　翻印必究·

凡购本社图书，如有印装错误，由本社读者服务部负责调换。
联系地址：北京阜外月坛北小街 2 号
电话：(010) 68022974　　邮编：100836

江西财经大学东亿论丛·第一辑
编委会

总主编：罗世华

编　委：罗良清　陶长琪　曹俊文　刘小瑜　魏和清
　　　　平卫英　刘小惠　徐　斌　杨头平　盛积良

总　序

江西财经大学统计学院源于 1923 年成立的江西省立商业学校会统科。统计学专业是学校传统优势专业，拥有包括学士、硕士（含专硕）、博士和博士后流动站的完整学科平台。数量经济学是我校应用经济学下的一个二级学科，拥有硕士、博士和博士后流动站等学科平台。

江西财经大学统计学科是全国规模较大、发展较快的统计学科之一。1978 年、1985 年统计专业分别取得本科、硕士办学权；1997 年、2001 年、2006 年统计学科连续三次被评为省级重点学科；2002 年统计学专业被评为江西省品牌专业；2006 年统计学硕士点被评为江西省示范性硕士点，是江西省第二批研究生教育创新基地。2011 年，江西财经大学统计学院成为我国首批江西省唯一的统计学一级学科博士点授予单位；2012 年，学院获批江西省首个统计学博士后流动站。2017 年，统计学科成功入选"江西省一流学科（成长学科）"；在教育部第四轮学科评估中被评为"A-"等级，进入全国前 10% 行列。目前，统计学科是江西省高校统计学科联盟盟主单位，已形成以研究生教育为先导、本科教育为主体、国际化合作办学为补充的发展格局。

我们推出这套系列丛书的目的，就是展现江西财经大学统计学院发展的突出成果，呈现统计学科的前沿理论和方法。之所以以"东亿"冠名，主要是以此感谢高素梅校友及所在的东亿国际传媒给予统计学院的大力支持，在学院发展的关键时期，高素梅校友义无反顾地为我们提供了无私的帮助。丛书崇尚学术精神,坚持专业视角,客观务实,兼具科学研究性、实际应用性、参考指导性,希望能给读者以启发和帮助。

丛书的研究成果或结论属个人或研究团队观点,不代表单位或官方结论。书中难免存在不足之处,恳请读者批评指正。

编委会

2019 年 6 月

Preface

In this book we consider how to measure the strength of various associations between two random variables, especially functional dependence. Several measures and their properties will be provided, then their estimations are discussed.

Among the measures of associations between random variables, Pearson's correlation coefficient, Spearman's ρ and Kendall's τ are the most prominent. They only measure linear or monotonic relationships, not suitable for nonlinear cases.

Here a measure for functional relationship will be provided. This functional relationship could be either linear or nonlinear. Joint distributions are often used in association analysis. One drawback of using joint distributions is that, even for the same type of association, for instance independence, joint distributions may be different. This is because joint distributions are not only affected by the dependence structure between random variables, but also affected by the variables themselves through their distributions. Instead of using joint distributions, our definition will make use of copula, which is a function that only describes association and will not be affected by the marginal distributions.

A measure of positive quadrant dependence, several measures of mutual complete dependence and functional dependence for discrete random variables will be provided. The measures of mutual complete dependence and functional dependence have two equivalent forms: one is constructed by measuring the "average" distance between conditional distributions and marginal distributions, and the other is constructed through subcopulas. Those measures are all ranged from 0

基于Copula的相关性测度
Measures of Association and Dependence Through Copulas

to 1, and the value of the measure tells us how strong the relationship is. Some properties of the measures are also obtained.

To apply those measures to sample data, several estimators are suggested. Unlike the joint distributions, which are directly observable, copulas are hidden dependence structures. This makes the task of proposing a parametric copula model non-trivial and is where a non parametric estimator can play a significant role. All estimators in this thesis are based on nonparametric kernel estimations, special emphasis are given to Beta kernel estimations.

Contents

1 Outline and Summary 1

   1.1 Introduction 1
   1.2 Outline 4

2 Statistical Modeling and Measurement of Association 5

   2.1 The concept of copulas 5
   2.2 Nonparametric estimations of copula 11
      2.2.1 An overview of empirical processes 11
      2.2.2 Nonparametric estimation via the empirical copula 13
      2.2.3 Functional delta-method and hadamard differentiability 13
      2.2.4 Weak convergence of the empirical copula process 15
      2.2.5 Nonparametric kernel estimations 15
      2.2.6 Bias and variance of kernel density estimator 18
      2.2.7 Optimal bandwith 20
   2.3 Measures of association and dependence 22
      2.3.1 Pearson's correlation coefficient 22
      2.3.2 Spearman's ρ and Kendall's τ 22
      2.3.3 The measure for mutual complete dependence 24
      2.3.4 The $*$ operator and the measure of mutual complete dependence 24

基于Copula的相关性测度
Measures of Association and Dependence Through Copulas

3 A Measure for Positive Quadrant Dependence ········ 27

4 Measures for Discrete MCD and Functional Dependence ······ 33
    4.1 The measure of MCD through conditional distributions ············ 34
    4.2 The measure of MCD through a subcopula ················ 39
    4.3 Comparison to Siburg and Stoimenov's measure of MCD ·········· 48
        4.3.1 Extension using E-process ················ 48
        4.3.2 Bilinear extension ················ 52
    4.4 Remarks on measures of dependence ················ 54
    4.5 Other measures ················ 56
        4.5.1 The measure μ_0^2 ················ 56
        4.5.2 The measure $\overline{\lambda}$ ················ 57
        4.5.3 Construction of the measure ················ 59
        4.5.4 Proofs of the construction of $\overline{\lambda}$ ················ 61

5 Nonparametric Estimation of the Measure of Functional Dependence ················ 70
    5.1 Nonparametric estimation through the density of copula ············ 71
        5.1.1 Estimating with pseudo-observations ················ 71
        5.1.2 Kernel estimation through copula density functions ······ 73
        5.1.3 Asymptotic behavior of the estimator of functional dependence ················ 76
    5.2 Nonparametric estimation of the measure of MCD via copula ······ 78
    5.3 Simulation results ················ 81

6 Implementation and Simulations ················ 88
    6.1 Choosing the evaluation grid ················ 88
    6.2 Simulation ················ 89

Contents

    6.3   Comparison of measures 92

7 Application 96

8 Discussion 98

References 99

Appendix 104

    A   List of Symbols 104
    B   Calculation of the Measure of PQD 106
    C   Beta Kernel Estimation 108
    D   Kernel Estimation 115
    E   FDM of variables in crime dataset 124

❶
Outline and Summary

1.1 Introduction

The study of association or dependence plays a central role in statistics. One of the important aspects of the study is how to measure the strength of various associations among random variables. The main purpose of this study is setting up copula-based measures of association, such as positive quadrant denpendence, mutual complete dependence, and functional dependences. Then the estimations of those measures are discussed. Further, simulations studies are carried out to investigate the performance of the estimators.

All measures introduced in this thesis are constructed via copula. Copulas capture the dependence structure among random variables, irrespective of their marginal distributions (see e. g. Joe (1997) and Nelsen (2006) for a detailed overview of copulas). Copula techniques are frequently applied in the quantitative finance literature, we mention Patton (2002), Embrechts et al. (2003), McNeil et al. (2005), Savu and Trede (2008), and Giacomini et al. (2009). Copulas allow marginal distributions and dependence structure to be modelled separately. One important character of copulas is that they are invariant under strictly increasing transformations, which make them an ideal tool to describe the dependence structure of random variables.

基于Copula的相关性测度
Measures of Association and Dependence Through Copulas

There are many different types of associations, such as positive quadrant dependence, stochastic increasing positive dependence, right-tail increasing and lefttail increasing, total positivity of order 2, tail dependence, those measures are summarized in Joe's study (1997). As outlined above, measures of multivariate association are naturally based on the copula of the underlying random vector. Various copula-based measures have been proposed in the literature. For example, Wolff (1980) introduces a class of multivariate measures of association which is based on the L_1 and L_∞ norms of the difference between the copula and the independent copula (see also Fernández-Fernández and González-Barrios (2004)). Other authors generalize existing bivariate measures of association to the multivariate case. For example, multivariate extensions of Spearman's rho are considered by Nelsen (1996) and Schmid and Schmidt (2006, 2007a, 2007b). Blomqvist's beta is generalized by Ubeda-Flores (2005) and Schmid and Schmidt (2007c), whereas a multivariate version of Gini's Gamma is proposed by Behboodian et al. (2007). Further, Joe (1990) and Nelsen (1996) discuss multivariate generalizations of Kendall's tau. A multivariate version of Spearman's footrule is considered by Genest et al. (2010). There are few papers concerning the discrete version of Kendall's τ and Spearman's ρ (see Nešlehová, Denuit (2005), Mesfioui (2005)).

Lehmann (1966) defined a joint distribution function, say $H(x, y)$ to be positively quadrant dependent (PQD) if $H(x, y) \geq H(x, \infty) H(\infty, y)$ for all x, y. Intuitively, X and Y are PQD if the probability that they are simultaneously small (or simultaneously large) is at least as great as it would be when they are independent. For example, a system may have components that are subject to the same set of stresses or shocks, or in which the failure of one component results in an increased load on the surviving components. In those situations, the short lifetime of component A may be accompanied with, or result in short lifetime of component B, so the dependence relationship between two lifetimes maybe described as PQD. Examples of positive quadrant dependence are ample in particular in insurance and finance. For the discussion about PQD in fi-

Outline and Summary

nance and actuarial sciences see Janic-Wróblewska, Kallenberg & Ledwina (2004) and Denuit & Scaillet (2004). For joint distributions which does not belong to PQD, a measure which can tell tell us how far those distributions are away from PQD will be defined.

Among the measures of associations between random variables, Pearson's correlation coefficient, Spearman's ρ and Kendall's τ are the most prominent. For reference, see Joe, Nelsen, Schmid and Schmidt, Taylor, and Wolff. But they only measure linear or monotonic relationships, not suitable for nonlinear cases. For example, when the relationship between two random variables is parabolic-shaped, none of the above measures will be applicable. We are going to define a measure for functional relationship between random variables, which could be either linear or nonlinear. The measure for functional relationship is derived from the measure of mutual complete dependence (MCD). Mutual complete dependence was first introduced by Lancaster (1982). Two random variables are mutually complete dependent when they have mutually functional relationship. This is also known as the strongest dependence. For independent relationship, two random variables are completely unpredictable of the other, while MCD, the strongest dependent relationship, correspond to complete predictability. Siburg (2010) constructed a measure of MCD for continuous random variables. Tasena et al. (2013) extended this measure to the multivariate case. In their paper, they proposed to measure the distance between two copulas by a modified Sobolev norm. For details of inner product and Sobolev norm on copulas space, I refer to Darsow et al. . But their measures are not applicable for discrete random variables, since both copulas and Sobolev norm are not suitable for discrete variables. So in this thesis, we will first define a measure of MCD for discrete random variables. Then considering that MCD is a mutual relationship, it is too strong in some cases. For example, when two random variables X and Y have relationship $Y=X^2$, it is still a very strong relationship, but not mutual. So we will also set up two measures for this kind of relationships.

1.2 Outline

In Chapter 2, the concept of copulas and several properties are introduced in Section 2.1. A summary of nonparametric estimations of copulas based on the empirical copula, kernel estimation and nonparametric bootstrap are presented in Section 2.2. After introducing several concepts of association in Section 2.3, Section 2.4 discusses the estimators of the measures of various types of association.

Chapter 3 discuss the measure of PQD and its properties. In Section 3.1, after introducing the concept of positive quadrant dependence (PQD), we discussed the problem of using Spearman's ρ as a measure of PQD. Then a measure of PQD and its analytical properties are discussed in Section 3.2. Finally, R code for calculating PQD measure of any copula is given in Section 3.3.

In Chapter 4, we defined a measure for discrete mutual complete dependence (MCD) and discussed its properties. First, $*$-operator in copula and the measure for continuous MCD are introduced in Setion 4.1. Then, Section 4.2 proposed a measure for discrete MCD and discussed its properties. Instead of defining a measure based on subcopula, another way is to extend the subcopula to copula, then use the measure for continuous MCD. In Section 4.3, we will show that this method has some drawbacks. Section 4.4 compares the measure for MCD with other measures of association, e.g. Spearman's ρ, Kendall's τ.

We provide several estimators of the measure for MCD based on empirical estimator and some types of kernel estimators in Chapter 5. Since pseudo observations are often used to estimate copulas, we will first discuss when pseudo observations will produce good estimator in Section 5.1. Then we derived the asymptotic distribution of the estimator based on Beta kernel. Section 5.2 provide another estimator of the measure through copulas. Finally, in Section 5.3, we report the simulation results.

❷
Statistical Modeling and Measurement of Association

2.1 The concept of copulas

The concept of copula was first introduced by Abe Sklar in 1959, but it didn't caught the attention of statisticians until recent years. Now people realized that copula is a very powerful tool to describe the relationships of random variables. It can describe not only the linear dependence, but also some other relationships, e.g. affiliation, concordance, PQD, MCD. Copulas play an important role for the following two reasons: Copulas are functions of dependence structure among random variables. Unlike joint distributions, they are not affected by marginal distributions of random variables. Further, They can be used to construct multivariate distribution functions by modeling marginal distribution functions and dependence structure separately. For example, using copulas, we can construct numerous joint distributions which are not normal but have normal marginals. Nowadays, it has been widely used in economics, especially in heavy tail analysis and risk analysis.

The copula is a multivariate distribution with univariate margins being uniform distribution $U(0, 1)$.

Definition 2.1 (Nelsen, 2006) A two-dimensional subcopula (or 2-subcopula) is a function C with the following properties:

基于Copula的相关性测度
Measures of Association and Dependence Through Copulas

a. $D(C) = D_1 \times D_2$, where $D(C)$ is the domain of C, D_1 and D_2 are subsets of $\mathbf{I} = [0, 1]$ containing 0 and 1

b. For every u in D_1 and every v in D_2,
$$C(u,0) = 0 = C(0,v) \quad \text{and} \quad C(u,1) = u \quad C(1,v) = v$$

c. For every u_1, u_2 in D_1, v_1, v_2 in D_2 such that $u_1 \leqslant u_2$ and $v_1 \leqslant v_2$,
$$C(u_2, v_2) - C(u_2, v_1) - C(u_1, v_2) + C(u_1, v_1) \geqslant 0$$

C becomes a **copula** when $D_1 = D_2 = \mathbf{I}$.

So a copula is also a bivariate distribution function restricted on the unit square $\mathbf{I}^2$. It is also a special case of subcopula. The above definition of 2-dimensional copulas can easily be extended to higher dimensions. The following theorem was given by Sklar (1959), it shows the relationship between joint distributions and copulas.

Theorem 2.1 Sklar's Theorem Let H be a joint distribution function with margins F and G. Then there exists a copula C such that for all x, y in $\overline{\mathbb{R}}$,
$$H(x, y) = C(F(x) \ G(y)) \tag{2-1}$$

where $\overline{\mathbb{R}}$ denote the extended real numbers. If F and G are continuous, then C is unique; otherwise, C is uniquely determined on Ran $F \times$ Ran G, where Ran F and Ran G denote the range of F and G respectively. Conversely, if C is a copula and F and G are distribution functions, then the function H defined by (2-1) is a joint distribution function with margins F and G.

Definition 2.2 Let F be a distribution function. Then a quasi-inverse of F is any function F^{-1} with domain $\mathbf{I}$ such that

a. if t is in Ran F, then $F^{-1}(t)$ is any number x in $\overline{\mathbb{R}}$ such that $F(x) = t$, i.e., for all t in Ran F,
$$F(F^{-1}(t)) = t$$

b. if t is not in Ran F, then
$$F^{-1}(t) = \inf\{x \mid F(x) \geqslant t\} = \sup\{x \mid F(x) \leqslant t\}$$

Using quasi-inverses of distribution functions, we have the following corollary to Theorem 2.1.

Corollary 2.1 Let H, F, G, and C be as in Theorem 2.1, and let F^{-1} and G^{-1} be quasi-inverses of F and G, respectively. Then for any (u, v) in the domain of C,

$$C(u, v) = H(F^{-1}(u), G^{-1}(v))$$

The following theorem shows us that unlike joint distributions, which are only bounded by $[0, 1]$, every copula is bounded by the so called Fréchet-Hoeffding boundaries.

Theorem 2.2 (Nelsen, 2006) Let C'' be a subcopula. Then for every (u, v) in $D(C'')$,

$$\max(u + v - 1, 0) \leq C''(u, v) \leq \min(u, v)$$

where $\min(u, v)$ is called Fréchet-Hoeffding upper bound, denoted by $C^+(u, v)$. And $\max(u + v - 1, 0)$ is called Fréchet-Hoeffding lower bound, denoted by $C^-(u, v)$.

Note that Theorem 2.2 can be extened to a n-dimensional copulas, where

$$C^+(u_1, \cdots, u_n) = \min\{u_1, \cdots, u_n\}$$

$$C^-(u_1, \cdots, u_n) = \max\{u_1 + \cdots + u_n - n + 1, 0\}$$

$C^+(u_1, \cdots, u_n)$ is a n-dimensional copula for all $n \geq 2$ and is known as the comonotonic copula. If the random vector X has copula $C^+(u_1, \cdots, u_n)$, each of the random variables $X_1, \cdots, X_n$ can (almost surely) be represented as strictly increasing function of any of the others. The copula C^+ is also said to describe perfect positive dependence. In contrast, C^- is only a copula for $n = 2$ and is called countermonotonic copula. It represents the copula of the bivariate random vector (X_1, X_2) if there exists a strictly decreasing relationship between X_1 and X_2. Here, the copula C^- describes the case of perfect negative dependence.

Another important copula is the independent copula $\Pi(u, v) = uv$. The following theorem will show us that independent copula describes independent structure of random variables.

基于Copula的相关性测度
Measures of Association and Dependence Through Copulas

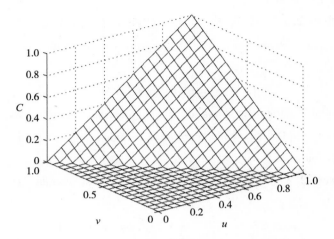

Figure 2-1 Fréchet-Hoeffding lower bound

Theorem 2.3 Nelsen (2006) Let X and Y be continuous random variables. Then X and Y are independent if and only if the corresponding copula $\Pi(u, v) = uv$.

An important property of copulas is that copulas are invariant under the strictly monotone transformations.

Theorem 2.4 (Schweizer, 1981) Let X and Y be continuous random variables with copula C_{XY}. If f and g are strictly increasing functions on Ran X and Ran Y, respectively, then $C_{f(X)g(Y)} = C_{XY}$. Thus C_{XY} is invariant under strictly increasing transformations of X and Y.

We mention a few more properties of copulas. These properties will be used in constructing the measure of MCD.

Proposition 2.1 Let C be a 2-dimensional copula, then the following properties hold:

a. C is increasing in each argument;
b. C is Lipschitz (and hence uniformly) continuous;
c. For $i \in 1, 2$, $\partial_i C$ exists a.e. on I^2 with $0 \leq \partial_i C(x, y) \leq 1$;

Statistical Modeling and Measurement of Association

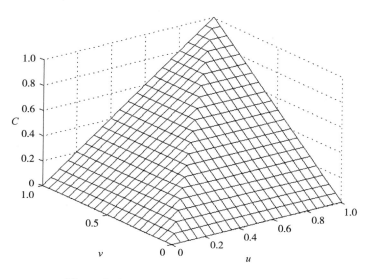

Figure 2-2 Fréchet-Hoeffding upper bound

d. The functions $t \to \partial_1 C(x, t)$ and $t \to \partial_2 C(t, y)$ are defined and increasing a. e. on **I**.

For the proof of this theorem, I refer to Nelsen (2006).

Example 2.1 Let (X, Y) be bivariate normal distribution. The joint distribution function is

$$H_\rho(x, y) = \int_{-\infty}^{x} \int_{-\infty}^{y} \frac{1}{2\pi\sqrt{1-\rho^2}} \exp\left\{-\frac{t^2 + s^2 - 2\rho ts}{2(1-\rho^2)}\right\} dt ds$$

where ρ is the correlation between X and Y. Note that the marginal distributions of X and Y are standard normal distributions Φ. Then the Gaussian copula is

$$C_\rho(u, v) = \int_{-\infty}^{\Phi^{-1}(u)} \int_{-\infty}^{\Phi^{-1}(v)} \frac{1}{2\pi\sqrt{1-\rho^2}} \exp\left\{-\frac{t^2 + s^2 - 2\rho ts}{2(1-\rho^2)}\right\} dt ds$$

Example 2.2 Let $H: I^2 \to I$ be the uniform distribution on the square I^2:

基于Copula的相关性测度
Measures of Association and Dependence Through Copulas

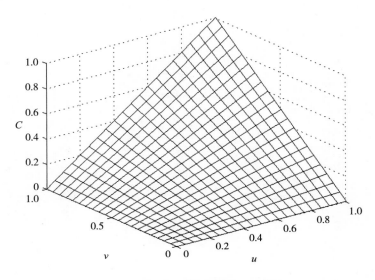

Figure 2-3 Independent copula

$$H(x,y) = \begin{cases} 0 & \text{if} \quad x \leqslant 0 \quad \text{or} \quad y \leqslant 0 \\ xy & \text{if} \quad 0 < x, y < 1 \\ 1 & \text{if} \quad x \geqslant 1 \quad \text{and} \quad y \geqslant 1 \end{cases}$$

Then the marginals F, G are both uniform distributions on I. So the corresponding copula is $C(u, v) = uv$.

Example 2.3 The joint distribution function H is

$$H(x,y) = \begin{cases} \frac{(x+1)(e^y-1)}{x+2e^y-1} & (x,y) \in [-1,1] \times [0,\infty] \\ 1 - e^{-y} & (x,y) \in (1,\infty] \times [0,\infty) \\ 0 & \text{elsewhere} \end{cases}$$

with margins F and G given by

$$F(x) = \begin{cases} 0 & x < -1 \\ (x+1)/2 & x \in [-1, 1] \\ 1 & x > 1 \end{cases}$$

and

$$G(y) = \begin{cases} 0 & y < 0 \\ 1 - e^{-y} & y \geqslant 0 \end{cases}$$

Pseudo-inverses of F and G are given by $F^{-1}(u) = 2u-1$ and $G^{-1}(v) = -\ln(1-v)$ for u, v in I, thus the copula C is

$$C(u, v) = \frac{uv}{u + v - uv}$$

2.2 Nonparametric estimations of copula

There are three general methods for the estimation of copulas: parametric, semiparametric and nonparametric methods. The parametric and semi-parametric estimation approaches are usually based on maximum-likelihood methods, we mention Genest and Rivest (1993), Genest et al. (1995), Joe and Xu (1996), Joe (2005), Chen and Fan (2006), and Kim et al. (2007). For an overview see also Malevergne and Sornette (2005), Chapter 5. In this thesis, we solely consider nonparametric estimation methods for which the joint and the marginal distribution functions are assumed to be unknown. Nonparametric estimation of copulas was first considered by Rüschendorf (1976) and Deheuvels (1979). Before discussing the empirical estimator of a copula, we give an overview of empirical processes.

2.2.1 An overview of empirical processes

Let T be an index set, ρ be a semimetric on T. We say that the semimetric

基于Copula的相关性测度
Measures of Association and Dependence Through Copulas

space (T, ρ) is totally bounded if for every $\epsilon > 0$, there exists a finite subset $T_k = t_1, \cdots, t_k \subset T$ such that for all $t \in T$, we have $\rho(s, t) \leq \epsilon$ for some $s \in T_k$.

A real-valued stochastic process $X_t, t \in T$, is a Gaussian process if all the finitedimensional distributions have a multivariate normal distribution. That is, for any choice of distinct values $t_1, \cdots, t_k \in T$, the random vector $X = (X_{t_1}, \cdots, X_{t_k})'$ has a multivariate normal distribution with mean vector $\mu = E(X)$ and covariance matrix $\Sigma = \text{cov}(X, X)$. The density function is given by

$$f_X(x) = (2\pi)^{-n/2} (\det\Sigma)^{-1/2} \exp\left(-\frac{1}{2}(x-\mu)'\Sigma^{-1}(x-\mu)\right)$$

A stochastic process is a collection of random variables $X(t), t \in T$ on the same probability space, indexed by an arbitrary index set T. An empirical process is a stochastic process based on a random sample. For example, for a random sample $X_1, \cdots, X_n$ of i.i.d. real random variables with distribution F. The empirical distribution function is

$$\mathbb{F}_n(t) = n^{-1} \sum_{i=1}^n \mathbb{1}\{X_i \leq t\}$$

where $\mathbb{1}$ is the indicator function and the index t is allowed to vary over $T = \mathbb{R}$, the real line.

More generally, we can consider a random sample $X_1, \cdots, X_n$ of independent draws from a probability measure P on an arbitrary sample space $\mathcal{X}$. We define the empirical measure to be $\mathbb{P}_n = n^{-1} \sum_{i=1}^n \delta_{X_i}$, where δ_x is the measure that assigns mass 1 at x and zero elsewhere. For a measurable function $f : \mathcal{X} \mapsto \mathbb{R}$, we denote $\mathbb{P}_n f = n^{-1} \sum_{i=1}^n f X_i$. For any class $\mathcal{F}$ of measurable functions $f : \mathcal{X} \mapsto \mathbb{R}$, an empirical process $\{\mathbb{P}_n f, f \in \mathbb{P}\}$ can be defined.

Setting $\mathcal{X} = \mathbb{P}$, we can re-express $\mathbb{F}_n$ as the empirical process $\{\mathbb{P}_n f, f \in \mathbb{P}\}$, where $\mathcal{F} = \{\mathbb{1}\{x \leq t\}, t \in \mathbb{R}\}$. Thus one can view the stochsatic process $\mathbb{F}_n$ as indexed by either $t \in \mathbb{R}$ or $f \in \mathcal{F}$. By the law of large numbers,

$$\mathbb{F}_n(t) \overset{as}{\to} F(t) \tag{2-2}$$

for each $t \in \mathbb{R}$, where $\overset{as}{\to}$ denotes almost sure convergence. A primary goal of empirical process is to study empirical processes as random functions over the associated index set. Each realization of one of these random functions is a sample path. To this end, Glivenko (1933) and Cantelli (1933) demonstrated that (2-2) could be strengthened to

$$\sup_{t \in \mathbb{R}} |\mathbb{F}_n(t) - F(t)| \xrightarrow{as} 0 \qquad (2-3)$$

Another way of saying this is that the sample paths of $\mathbb{F}_n$ get uniformly closer to F as $n \to \infty$. Returning to general empirical processes, a class $\mathcal{F}$ of measurable functions $f : \chi \mapsto \mathbb{R}$, is said to be a P-Glivenko-Cantelli class if

$$\sup_{f \in \mathcal{F}} |\mathbb{P}_n f - Pf| \xrightarrow{as} 0 \qquad (2-4)$$

2.2.2 Nonparametric estimation via the empirical copula

Consider a random sample $(X_1, Y_1), (X_2, Y_2), \cdots (X_n, Y_n)$ of i.i.d. from joint distribution H with continuous marginal distribution functions $F_X(x)$ and $G_Y(y)$. we construct the empirical distribution function

$$\mathbb{H}_n(x, y) = \frac{1}{n} \sum_{i=1}^{n} \mathbb{1}_{\{X_i \leqslant x, Y_i \leqslant y\}} \quad -\infty < x, y < +\infty$$

Empirical copula estimator:

$$\hat{C}_n(u, v) = \frac{1}{n} \sum_{i=1}^{n} \mathbb{1}_{\{\hat{U}_i \leqslant u, \hat{V}_i \leqslant v\}} = \frac{1}{n} \sum_{i=1}^{n} \mathbb{1}_{\{\hat{U}_i \leqslant u\}} \mathbb{1}_{\{\hat{V}_i \leqslant v\}} \qquad (2-5)$$

with $\hat{U}_i = F_n(X_i)$, $\hat{V}_i = G_n(Y_i)$, where F_n and G_n are the empirical cumulative distribution functions of the marginals, and where $\mathbb{1}_A$ denotes the indicator of a set A.

Weak convergence of the empirical copula process $\sqrt{n}(\hat{C}_n - C)$ can be estimated using the functional delta-method, which will be introduced next.

2.2.3 Functional delta-method and hadamard differentiability

The classical delta-method represents an important technique for deducing the asymptotic distribution of a sequence of transformed random vectors from the asymptotic behavior of the underlying sequence. Let $\ell^{\infty}(I^n)$ be the space of the

基于Copula的相关性测度
Measures of Association and Dependence Through Copulas

collection of all uniformly bounded real-valued functions defined on I^n, equipped with the uniform metric μ defined as

$$\mu(f_1, f_2) = \sup_{t \in I^n} |f_1(t) - f_2(t)|, \quad f_1, f_2 \in \ell^\infty(I^n) \tag{2-6}$$

Assuming that every sample path $\mathbf{u} \mapsto (\hat{C}_n(\mathbf{u}))(\omega)$ of the empirical copula $\hat{C}_n$ is a bounded function on I^n, the empirical copula can be viewed as the following map:

$$\hat{C}_n : \Omega \times I^n \mapsto \ell^\infty(I^n)$$

i. e. as a random map taking values in the function space $\ell^\infty(I^n)$. In this context, a general version of the delta-method addressing the weak convergence of stochastic processes is needed. This functional delta-method is based on the notion of Hadamard differentiability. Let $\mathbb{D}$ and $\mathbb{E}$ be two metrizable, topological spaces (i. e. vector spaces for which addition and scalar multiplication are continuous operations).

Definition 2.3 A map $\phi : \mathbb{D}_\phi \subset \mathbb{D} \mapsto \mathbb{E}$ is called Hadamard differentiable at $\theta \in \mathbb{D}_\phi$ if there is a continuous linear map $\phi'_\theta : \mathbb{D} \mapsto \mathbb{E}$ such that

$$\frac{\phi(\theta + t_n h_n) - \phi(\theta)}{t_n} \to \phi'_\theta(h), \quad n \to \infty$$

for all converging sequences $t_n \to 0$ and $h_n \to h$ such that $\theta + t_n h_n \in \mathbb{D}_\phi$ for every n.

This definition may be refined to Hadamard differentiable tangentially to a set $\mathbb{D}_0 \subset \mathbb{D}$ by requiring that every $h_n \to h$ in the definition has $h \in \mathbb{D}_0$.

The function ϕ'_θ is called the Hadamard derivative of the map ϕ at $\theta \in \mathbb{D}_\phi$. Thereby, the set $\mathbb{D}_\phi$ can be any arbitrary subset of $\mathbb{D}$. For $h \in \mathbb{D}_0$, the derivative $\phi'_\theta(h)$ represents a first order Taylor approximation evaluated at the point θ in direction h. For maps $\phi : \mathbb{D}_\phi \subset \mathbb{R}^m \mapsto \mathbb{R}^n$, Hadamard differentiability is equivalent to the usual type of differentiability.

Theorem 2.5 (Functional delta-method) Let $\mathbb{D}$ and $\mathbb{E}$ be normed spaces. Let $\phi : \mathbb{D}_\phi \subset \mathbb{D} \mapsto \mathbb{E}$ be Hadamard-differentiable at θ tangentially to $\mathbb{D}_0$. Let $X_n : \Omega_n \mapsto \mathbb{D}_\phi$ be maps with $r_n(X_n - \theta) \xrightarrow{w} X$ for some sequence of constants $r_n \to \infty$, where X is separable and takes its values in $\mathbb{D}_0$. Then $r_n(\phi(X_n) - \phi$

$(\theta))\xrightarrow{w}\phi'_\theta(X).$

2.2.4 Weak convergence of the empirical copula process

The copula C of the distribution function H can be represented as a map ϕ: $D(\mathbb{R}^2) \mapsto \ell^\infty(I^2)$ of H via

$$C(u, v) = \phi(H)(u, v) = H(F^{-1}(u), G^{-1}1(v)) \quad (2-7)$$

Lemma 2.1 Fix $0 < p < q < 1$, and suppose that H is a distribution function on $\mathbb{R}^2$ with marginal distribution functions F and G that are continuously differentiable on the intervals $[F^{-1}(p) -\epsilon, F^{-1}(q) +\epsilon]$ and $[G^{-1}(p) -\epsilon, G^{-1}(q) +\epsilon]$ with positive derivative f and g, respectively, for some >0. Furthermore, assume that $\partial H/\partial x$ and $\partial H/\partial y$ exist and are continuous on the product of these intervals, then the map $\phi: D(\mathbb{R}^2) \mapsto \ell^\infty([p, q]^2)$ is Hadamard-differentiable at H tangentially to $C(\mathbb{R}^2)$. The derivative is given by

$$\phi'_H(h)(u,v) = h(F^{-1}(u), G^{-1}(v)) - \frac{\partial H}{\partial x}(F^{-1}(u), G^{-1}(v))\frac{h(F^{-1}(u), \infty)}{f(F^{-1}(u))} - \frac{\partial H}{\partial y}(F^{-1}(u), G^{-1}(v))\frac{h(\infty, G^{-1}(v))}{g(G^{-1}(v))}$$

Theorem 2.6 Schmid (2007) Let F be a continuous d-dimensional distribution function with copula C. $\hat{C}_n$ is defined in (2-5). Under the additional assumption that the i-th partial derivatives $\partial_i C(u, v)$ exist and are continuous for $i = 1, 2$, we have

$$\sqrt{n}\{\hat{C}_n(u,v) - C(u,v)\} \xrightarrow{w} \mathbb{G}_c(u,v)$$

Weak convergence takes place in $\ell^\infty(I^2)$ and

$$\mathbb{G}_c(u,v) = \mathbb{B}_c(u,v) - \partial_1 C(u,v)\mathbb{B}_c(u,1) - \partial_2 C(u,v)\mathbb{B}_c(1,v)$$

where $\mathbb{B}c(u, v)$ is a Brownian bridge on I^2 with covariance function

$$E[\mathbb{B}_c(u,v), \mathbb{B}_c(u',v')] = C(u \wedge u', v \wedge v') - C(u,v)C(u',v')$$

for each $0 \leqslant u, u', v, v' \leqslant 1$.

2.2.5 Nonparametric kernel estimations

Definition 2.4 The kernel is a continuous, bounded and symmetric real

Measures of Association and Dependence Through Copulas

function K such that it integrates to one, i. e.

$$\int_{-\infty}^{+\infty} K(u)du = 1$$

In many applications nonnegative kernels are used, so that $K(u) \geqslant 0$ for any u, so that K is a density function of a symmetric distribution with zero mean.

Some typical kernels:

$$\text{Uniform } K(u) = \frac{1}{2} \mathbb{1}_{\{|u| \leqslant 1\}};$$

Triangular kernel:

$$K(u) = (1-|u|) \mathbb{1}_{\{|u| \leqslant 1\}}$$

Normal kernel:

$$K(u) = \frac{1}{\sqrt{2\pi}} \exp\left(-\frac{u^2}{2}\right)$$

Epanechnikov kernel:

$$K(u) = \begin{cases} \frac{3}{4}(1-u^2) & |u| \leqslant 1 \\ 0 & |u| > 1 \end{cases}$$

Biweight or Quartic kernel:

$$K(u) = \begin{cases} \frac{15}{16}(1-u^2)^2 & |u| \leqslant 1 \\ 0 & |u| > 1 \end{cases}$$

For a bandwidth h, define

$$K_h(u) = \frac{1}{h} K\left(\frac{u}{h}\right)$$

Definition 2.5 Kernel density estimator of density function $f(x)$ is

$$\hat{f}(x) = \frac{1}{n} \sum_{i=1}^{n} K_h(x - X_i)$$

$$= \frac{1}{nh} \sum_{i=1}^{n} K\left(\frac{x - X_i}{h}\right)$$

Just for a moment, let's treat our sample $\{X_i\}$ as fixed and consider a random variable with density $\hat{f}$. Then its mean is

Statistical Modeling and Measurement of Association

$$\int_{-\infty}^{+\infty} x\hat{f}(x)dx = \int_{-\infty}^{+\infty} \frac{1}{nh}\sum_{i=1}^{n} xK\left(\frac{x-X_i}{h}\right)dx$$

$$= \frac{1}{n}\sum_{i=1}^{n}\int_{-\infty}^{+\infty}(uh+X_i)K(u)du$$

$$= \frac{1}{n}\sum_{i=1}^{n}\left[\int_{-\infty}^{+\infty}uhK(u)du + \int_{-\infty}^{+\infty}X_iK(u)du\right]$$

$$= \frac{1}{n}\sum_{i=1}^{n}h\int_{-\infty}^{+\infty}uK(u)du + \frac{1}{n}\sum_{i=1}^{n}X_i\int_{-\infty}^{+\infty}K(u)du$$

$$= \frac{1}{n}\sum_{i=1}^{n}X_i$$

So, mean of a random variable with density $\hat{f}$ is just a sample mean (and as such it does not depend on the choice of Kernel K and bandwidth h).

Second moment of $\hat{f}$:

$$\int_{-\infty}^{+\infty} x^2\hat{f}(x)dx = \int_{-\infty}^{+\infty} \frac{1}{nh}\sum_{i=1}^{n} x^2K\left(\frac{x-X_i}{h}\right)dx$$

$$= \frac{1}{n}\sum_{i=1}^{n}\int_{-\infty}^{+\infty}(uh+X_i)^2K(u)du$$

$$= \frac{1}{n}\sum_{i=1}^{n}\int_{-\infty}^{+\infty}u^2h^2K(u)du + \frac{1}{n}\sum_{i=1}^{n}\int_{-\infty}^{+\infty}X_i^2K(u)du +$$

$$\frac{2}{n}\sum_{i=1}^{n}\int_{-\infty}^{+\infty}uhX_iK(u)du$$

$$= \frac{1}{n}\sum_{i=1}^{n}h^2\int_{-\infty}^{+\infty}u^2K(u)du + \frac{1}{n}\sum_{i=1}^{n}X_i^2 +$$

$$\frac{2}{n}\sum_{i=1}^{n}hX_i\int_{-\infty}^{+\infty}uK(u)du$$

$$= \frac{1}{n}\sum_{i=1}^{n}X_i + \sigma_K^2 h^2$$

where

$$\sigma_K^2 = \int_{-\infty}^{+\infty} u^2 K(u)du$$

Therefore, variance of $\hat{f}$ is

基于Copula的相关性测度
Measures of Association and Dependence Through Copulas

$$\int_{-\infty}^{+\infty} x^2 \hat{f}(x) dx - \left(\int_{-\infty}^{+\infty} x\hat{f}(x) dx\right)^2 = \frac{1}{n}\sum_{i=1}^n X_i + \sigma_K^2 h^2 - \left(\frac{1}{n}\sum_{i=1}^n X_i\right)^2$$
$$= \sigma_K^2 h^2 + \hat{\sigma}_X^2$$

Variance of a random variable with density $\hat{f}$ is therefore sample variance $\hat{\sigma}_X^2$ plus a positive term $\sigma_K^2 h^2$.

2.2.6 Bias and variance of kernel density estimator

Now we treat the sample $\{X_i\}$ as random varialbes, so that $\hat{f}(x)$ is a random variable. Then its expected value is

$$E[\hat{f}(x)] = E\left[\frac{1}{nh}\sum_{i=1}^n K\left(\frac{x-X_i}{h}\right)\right]$$
$$= E\left[\frac{1}{h} K\left(\frac{x-X_i}{h}\right)\right]$$
$$= \frac{1}{h}\int_{-\infty}^{+\infty} K\left(\frac{x-z}{h}\right) f(z) dz$$
$$= \int_{-\infty}^{+\infty} K(u) f(x-hu) du$$

Therefore, $E[\hat{f}(x)] = f(x)$, so that kernel density estimator is biased. Also, $E[\hat{f}(x)] \to f(x)$ as $h \to 0$ if f is continuous and bounden from above, so kernel density estimator is asymptotically unbiased.

If density f is smooth enough around x, then we can use its Taylor series expansion around x to evaluate small sample bias of kernel estimator:

$$f(x-hu) = f(x) - f'(x)hu + \frac{1}{2}f''(x)h^2 u^2 + O(h^3)$$

so that

$$E[\hat{f}(x)] = \int_{-\infty}^{+\infty} K(u) f(x-hu) du$$
$$= \int_{-\infty}^{+\infty} K(u)\left(f(x) - f'(x)hu + \frac{1}{2}f''(x)h^2 u^2 + O(h^3)\right) du$$
$$= f(x) + \frac{h^2}{2} f''(x)\sigma_K^2 + O(h^3)$$

So, if we use a symmetric kernel, then the bias is of the order h^2.

Statistical Modeling and Measurement of Association

Variance of the kernel estimator is

$$V(x) = V\left[\frac{1}{n}\sum_{i=1}^{n}K_h(x-X_i)\right]$$

$$= \frac{1}{n}V(K_h(x-X_i))$$

$$= \frac{1}{n}(E[K_h(x-X_i)]^2 - (E[K_h(x-X_i)])^2)$$

$$= \frac{1}{n}\int_{-\infty}^{+\infty}\left[\frac{1}{h}K\left(\frac{x-z}{h}\right)\right]^2 f(z)dz - \frac{1}{n}\left(\int_{-\infty}^{+\infty}\frac{1}{h}K\left(\frac{x-z}{h}\right)f(z)dz\right)^2$$

(2-8)

Let $\dfrac{x-z}{h}=u$, then $z=x-hu$. Then the first term of (2-8) is

$$\frac{1}{n}\int_{-\infty}^{+\infty}\left[\frac{1}{h}K\left(\frac{x-z}{h}\right)\right]^2 f(z)dz$$

$$= \frac{1}{n}\int_{+\infty}^{-\infty}\left[\frac{1}{h}K(u)\right]^2 f(x-hu)(-h)du$$

$$= \frac{1}{n}\int_{-\infty}^{+\infty}\frac{1}{h}K(u)^2 f(x-hu)du$$

$$= \frac{1}{n}\int_{-\infty}^{+\infty}\frac{1}{h}K(u)^2\left(f(x)-f'(x)hu+\frac{1}{2}f''(x)h^2u^2+O(h^3)\right)du$$

$$= \frac{1}{n}\int_{-\infty}^{+\infty}\frac{1}{h}K(u)^2(f(x)-f'(x)hu+O(h^2))du$$

The second term of (2-8) is

$$\frac{1}{n}\left(\int_{-\infty}^{+\infty}\frac{1}{h}K\left(\frac{x-z}{h}\right)f(z)dz\right)^2$$

$$= \frac{1}{n}\left(\int_{-\infty}^{+\infty}K(u)f(x-hu)du\right)^2$$

$$= \frac{1}{n}\left(f(x)+\frac{h^2}{2}f''(x)\sigma_K^2+O(h^3)\right)^2$$

$$= \frac{1}{n}(f(x)+O(h^2))^2$$

So,

基于Copula的相关性测度
Measures of Association and Dependence Through Copulas

$$V(x) = \frac{1}{n}\int_{-\infty}^{+\infty} \frac{1}{h}K(u)^2(f(x) - f'(x)hu + O(h^2))\mathrm{d}u + \frac{1}{n}(f(x) + O(h^2))^2$$

$$= \frac{1}{nh}\int_{-\infty}^{+\infty} K(u)^2(f(x) - f'(x)hu + O(h^2))\mathrm{d}u + \frac{1}{n}(f(x) + O(h^2))^2$$

$$= \frac{1}{nh}\left(\int_{-\infty}^{+\infty} K(u)^2 f(x)\mathrm{d}u - \int_{-\infty}^{+\infty} K(u)^2 f'(x)hu\mathrm{d}u\right) +$$

$$\frac{1}{nh}\int_{-\infty}^{+\infty} K(u)^2 O(h^2)\mathrm{d}u + \frac{1}{n}(f(x) + O(h^2))^2$$

$$= \frac{1}{nh}\left(f(x)\int_{-\infty}^{+\infty} K(u)^2\mathrm{d}u - f'(x)\int_{-\infty}^{+\infty} K(u)^2 hu\mathrm{d}u\right)$$

$$+ \frac{1}{nh}O(h^2)\int_{-\infty}^{+\infty} K(u)^2\mathrm{d}u + \frac{1}{n}(f(x) + O(h^2))^2$$

$$= \frac{1}{nh}f(x)C_K + o\left(\frac{1}{nh}\right)$$

where

$$C_K = \int_{-\infty}^{+\infty} K(u)^2\mathrm{d}u$$

That is, the variance of $\hat{f}(x)$ is of the order (nh) (compare with n for variance of a parametric estimator).

2.2.7 Optimal bandwith

The researcher has a flexibility of choosing kernel K and bndwidth h. We noticed above that the choice of a kernel does not play a significant role, however the choice of the bandwith h does (look at the asymptotic approximation fo the variance of kernel density estimator). To make a choice of h, we need some criterion of optimality. Two most popular criteria are minimization of Integraded Squared Error (ISE) and Mean Integrated Squared Error (MISE), where

$$\mathrm{ISE}(h) = \int_{-\infty}^{+\infty} [\hat{f}(x) - f(x)]^2 \mathrm{d}x$$

$$\mathrm{MISE}(h) = E(\mathrm{ISE}(h)) = E\left(\int_{-\infty}^{+\infty} [\hat{f}(x) - f(x)]^2 \mathrm{d}x\right)$$

Both ISE and MISE are some measures of how good our kernel density estimator approximates true density function.

Finally, for a fixed x, we can define Mean Squared Error (MSE) as

$$\mathrm{MSE}(x;h) = E(\hat{f}(x) - f(x))^2$$
$$= [E(\hat{f}(x) - f(x))]^2 + V(\hat{f}(x) - f(x))$$
$$= [E(\hat{f}(x) - f(x))]^2 + V(\hat{f}(x))$$
$$= (E[\hat{f}(x)] - f(x))^2 + E(\hat{f}(x) - E[\hat{f}(x)])^2$$

Since we can change the order of integration, we have

$$\mathrm{MISE}(h) = E\left(\int_{-\infty}^{+\infty} [\hat{f}(x) - f(x)]^2 dx\right)$$
$$= \int_{-\infty}^{+\infty} (E(\hat{f}(x) - f(x))^2) dx$$
$$= \int_{-\infty}^{+\infty} \mathrm{MSE}(x;h) dx$$

so that MISE equals to Integrated MSE.

Finally, let $B(x) = E[\hat{f}(x)] - f(x)$ be the bias of kernel density estimator.
Then

$$\mathrm{MSE}(x;h) = [E(\hat{f}(x) - f(x))]^2 + V(\hat{f}(x) - f(x))$$
$$= (E[\hat{f}(x)] - f(x))^2 + V(\hat{f}(x))$$
$$= B(x)^2 + V(\hat{f}(x))$$
$$\mathrm{MISE}(h) = \int B(x)^2 dx + \int V(\hat{f}(x)) dx$$

We would want to find a way to pick h such that it minimizes MISE. However, the exact expression for MISE is hard to obtain (since it depends on the unknown DGP f). We can use asymptotic approximations to bias and variance of kernel density estimator derived above to approximate MISE with Asymptotic MISE (AMISE):

$$\mathrm{AMISE}(h) = \frac{h^4}{4}\sigma_K^2 \int f''(x)^2 dx + \frac{1}{nh} C_K$$

Optimal bandwidth h_{opt} is the one that minimizes AMISE, so that
$$h_{\text{opt}} = an^{-1/5}$$
where a is a constant that depends on true density f and kernel K.

2.3 Measures of Association and Dependence

2.3.1 Pearson's correlation coefficient

There are a variety of ways to discuss and to measure dependence. We will discuss a few here. The first one is Pearson's correlation coefficient. It is in general for spherical and elliptical distributions. The Pearson's linear correlation coefficient r for random vector (X_1, X_2) with joint distribution function H and marginals F_1 and F_2 is defined by

$$r(X_1, X_2) = \frac{E(X_1 - E[X_1])(X_2 - E[X_2])}{\sqrt{Var[X_1]Var[X_2]}}$$

Pearson's linear correlation can be expressed in copula form:

$$r(X_1, X_2) = \frac{1}{\sqrt{Var[X_1]Var[X_2]}} \int_{(u_1,u_2)\in \mathbf{I}^2} [C(u_1, u_2) - u_1 u_2] dF_1^{-1}(u_1) dF_2^{-1}(u_2)$$

2.3.2 Spearman's ρ and Kendall's τ

The popular dependence measures such as Spearman's ρ, Kendall's τ and Gini's γ are measures of concordance, a concept introduced by Scarsini (1984), and discussed e.g., by Joe (1997) and Nelsen (1991, 2002, 2006). Spearman's and Kendall's rank correlations measure the degree of concordance between X and Y, therefore, they are called measures of concordance. Informally, two random variables are concordant if "large" values of one tend to be associated with "large" values of the other and "small" values of one with "small" values of the other. Definitions for multivariate measures of concordance are given in Dolati and Úbeda-Flores and Taylor.

Statistical Modeling and Measurement of Association

First, we give the definition of concordance and discordance.

Definition 2. 6 Let (x_i, y_i) and (x_j, y_j) denote two observations from a vector (X, Y) of continuous random variables.

(x_i, y_i) and (x_j, y_j) are said to be concordant if $(x_i - x_j)(y_i - y_j) > 0$;
(x_i, y_i) and (x_j, y_j) are said to be discordant if $(x_i - x_j)(y_i - y_j) < 0$.

Definition 2. 7 If C_1 and C_2 are copulas, we say that C_1 is smaller than C_2 (or C_2 is larger than C_1), and write $C_1 \prec C_2$ (or $C_2 \succ C_1$) if $C_1(u, v) \leq C_2(u, v)$ for all u, v in I.

Scarsini (1984) gave the axiomatic definition of a measure of concordance.

Definition 2. 8 A numeric measure κ of association between two continuous random variables X and Y whose copula is C is a measure of concordance if it satisfies the following properties (we write $\kappa_{X,Y}$ or k_C when convenient):

a. κ is defined for every pair X, Y of continuous random variables;

b. $-1 \leq \kappa_{X,Y} \leq 1$, $\kappa_{X,X} = 1$, and $\kappa_{X,-X} = -1$;

c. $\kappa_{X,Y} = \kappa_{Y,X}$;

d. if X and Y are independent, then $\kappa_{X,Y} = 0$;

e. $\kappa_{-X,Y} = \kappa_{X,-Y} = -\kappa_{X,Y}$;

f. if C_1 and C_2 are copulas such that $C_1 \prec C_2$, then $\kappa_{C_1} \leq \kappa_{C_2}$;

g. if (X_n, Y_n) is a sequence of continuous random variables with copulas C_n, and if C_n converges pointwise to C, then $\lim_{n \to \infty} \kappa_{C_n} = \kappa_C$.

Definition 2. 9 Kendall's τ is defined as the probability of concordance minus the probability of discordance:

$$\tau = \tau_{X,Y} = P[(X_1 - X_2)(Y_1 - Y_2) > 0] - P[(X_1 - X_2)(Y_1 - Y_2) < 0]$$

Definition 2. 10 Let (X_1, Y_1), (X_2, Y_2), and (X_3, Y_3) be three independent random vectors with a common joint distribution function H (whose margins are again F and G) and copula C. $\rho_{X,Y}$ of Spearman's ρ is defined to be proportional to the probability of concordance minus the probability of discordance for the two vectors (X_1, Y_1) and (X_2, Y_3) i. e.

$$\rho_{X,Y} = 3(P[(X_1 - X_2)(Y_1 - Y_3) > 0] - P[(X_1 - X_2)(Y_1 - Y_3) < 0])$$

基于Copula的相关性测度
Measures of Association and Dependence Through Copulas

The following two theorems give us the copula form of Spearman's ρ and Kendall's τ.

Theorem 2.7 Let X and Y be continuous random variables whose copula is C. Then Kendall's τ for X and Y (which we will denote by either $\tau_{X,Y}$, or τ_C) is given by

$$\tau_{X,Y} = \tau_C = 4\int_{\mathbf{I}^2} C(u,v)\mathrm{d}C(u,v) - 1$$

Theorem 2.8 Let X and Y be continuous random variables whose copula is C. Then the popular version of Spearman's ρ for X and Y (which we will denoteby either $\rho_{X,Y}$, or ρ_C) is given by

$$\rho_{X,Y} = \rho_C = 12\int_{\mathbf{I}^2} uv\mathrm{d}C(u,v) - 3 = 12\int_{\mathbf{I}^2} C(u,v)\mathrm{d}u\mathrm{d}v - 3$$

2.3.3 The measure for mutual complete dependence

There is another type of measure for dependence called mutual complete dependence, see Lancaster, which is different from the measures given above. Let X and Y be two random variables. Y is said to be completely depend on X if there exists a Borel function f such that

$$P(Y = f(X)) = 1$$

X and Y are called mutually complete dependent (MCD) if Y is completely depend on X and X is also completely depend on Y. Note that Spearman's ρ and Kendall's τ are measures for concordance, while MCD is a measure for functional relationships between X and Y.

2.3.4 The * operator and the measure of mutual complete dependence

The set $\mathbb{C}$ can be equipped with the *-multiplication, defined by

$$(A * B)(x,y) = \int_0^1 \partial_2 A(x,t)\partial_1 B(t,y)\mathrm{d}t$$

After easy calculation, we will find

$$C^+ * C = C * C^+ = C$$

Statistical Modeling and Measurement of Association

$$\Pi * C = C * \Pi = \Pi$$

Theorem 2.9 For all $A, B \in \mathbb{C}$, $A * B \in \mathbb{C}$.

Define scalar product for copulas. For $A, B \in \mathbb{C}$,

$$\langle A, B \rangle = \int_{I^2} \nabla A \cdot \nabla B \, d\lambda \qquad (2-9)$$

where ∇ denotes the gradient of the copula.

Theorem 2.10 Let $A, B \in \mathbb{C}$. Then

$$\frac{1}{2} \leq \langle A, B \rangle \leq 1$$

where both bounds are sharp

The norm of copula C induced by inner product in (2-9) is given by:

$$\|C\|^2 = \int_0^1 \int_0^1 \left[\left(\frac{\partial C}{\partial u}\right)^2 + \left(\frac{\partial C}{\partial v}\right)^2 \right] du\,dv \qquad (2-10)$$

Definition 2.11 For any copula C, the copula C^T defined by

$$C^T(u,v) = c(v,u)$$

for all $(u, v) \in I^2$, is called the transposed copula of C.

Theorem 2.11 The Sobolev norm for copulas satisfies $\|C\|^2 \in [2/3, 1]$ for all $C \in \mathbb{C}$. Moreover, the following properties hold:

a. $\|C\|^2 = 2/3$ if and only if X and Y are independent;

b. $\|C\|^2 = 1$ if and only if X and Y are MCD.

Corollary 2.2 The following are equivalent:

a. X and Y are MCD;

b. $\|C\|^2 = 1$;

c. $\partial_1 C, \partial_2 C \in \{0, 1\}$ a. e. ;

d. C is invertible, i. e. , $C * C^T = C^T * C = C^+$.

Definition 2.12 Given two continuous random variables X, Y with copula C, we define

$$\mu(X,Y) = (3\|C\|^2 - 2)^{1/2}$$

$\mu(X, Y)$ can be interpreted as a normalized Sobolev distance of C from the independent copula Π:

基于Copula的相关性测度
Measures of Association and Dependence Through Copulas

$$\mu(X,Y) = \sqrt{3}\|C - \Pi\| = \frac{\|C - \Pi\|}{\|C_m - \Pi\|}$$

where C_m is a MCD copula.

Theorem 2.12 The quantity $\mu(X, Y)$ has the following properties:

a. $\mu(X, Y)$ is defined for any X and Y;

b. $\mu(X, Y) = \mu(Y, X)$;

c. $0 \leqslant \mu(X, Y) \leqslant 1$;

d. $\mu(X, Y) = 0$ if and only if X and Y are independent;

e. $\mu(X, Y) = 1$ if and only if X and Y are MCD;

f. $\mu(X, Y) \in [1/2, 1]$ if Y is completely depend on X;

g. If f and g are a.s. strictly monotone functions on Range(X) and Range(Y), respectively, then $\mu(f(X), g(Y)) = \mu(X, Y)$;

h. If $\{(X_n, Y_n)\}$ is a sequence of pairs of continuous random variables with copula C_n, and if $\lim_{n\to\infty} \|C_n - C\| = 0$, then $\lim_{n\to\infty} \mu(X_n, Y_n) = \mu(X, Y)$.

❸
A Measure for Positive Quadrant Dependence

Positive quadrant dependence (PQD) is one sort of relationship between random variables. As mentioned in the Introduction, examples of positive quadrant dependence are ample in particular in insurance and finance. We are interested in this question: for a joint distributions which is not PQD, can we define a "distance" from this distribution to the PQD set? The whole copula set or all possible relationships can be classified as three sets: positive quadrant dependence, negative quadrant dependence and others. In this section, we are going to define a measure for any joint distribution, and this measure will tell us the "distance" from this distribution to PQD set.

Definition 3.1 Let X and Y be random variables. X and Y are positively quadrant dependent (PQD) if for all (x, y) in $\mathbb{R}^2$,
$$P[X \leqslant x, Y \leqslant y] \geqslant P[X \leqslant x]P[Y \leqslant y]$$
or equivalently,
$$P[X > x, Y > y] \geqslant P[X > x]P[Y > y]$$

Definition 3.2 Let X and Y be random variables. X and Y are positively quadrant dependence (PQD) if
$$H(x, y) \geqslant F(x)G(y) \tag{3-1}$$
for all $(x, y) \in \mathbb{R}^2$, which is equivalent to
$$C(u, v) \geqslant uv \tag{3-2}$$

The PQD is a "global" property, since (3-2) must be hold for all points in I^2. A copula satisfies (3-2) for some $(u, v) \in [0, 1]^2$ is called locally PQD.

基于Copula的相关性测度
Measures of Association and Dependence Through Copulas

Recall that one of the form of Spearman's ρ is

$$\rho_C = 12 \int\int_{\mathbf{I}^2} [C(u,v) - uv] du dv$$

hence ρ_C can be interpreted as an "average" measure of quadrant dependence. But one problem of this measure is: it may not be able to distinguish local PQD and PQD.

Let's consider two copulas.

Example 3.1 Let t be in $[0, 1)$, and let C be the function from $\mathbb{I}^2$ into $\mathbb{I}$ given by

$$C_t(u,v) = \begin{cases} \max(u+v-1, t) & (u,v) \in [t,1]^2 \\ \min(u,v) & otherwise \end{cases} \quad (3-3)$$

If $t \neq 0$, (3-3) is a locally PQD. Spearman's $\rho_C = 0.984$, for $t = 0.8$.

Example 3.2 Consider Ali-Mikhail-Haq Copula (see Figure 5),

$$C(u,v) = \frac{uv}{1 - \theta(1-u)(1-v)}$$

We know it is PQD for $\theta > 0$. Spearman's $\rho_C = 0.034$, for $\theta = 0.1$.

Since Spearman's ρ is an "average" measure, PQD copula may have small value of ρ, while locally PQD copula may have big ρ. We want to a measure which can tells us how close a given copula to the PQD set.

Let $\mathcal{C}$ be the set of all copulas.

Definition 3.3 For any copula $C(u,v) \in \mathcal{C}$, define $m(C): \mathcal{C} \to \mathbf{I}$ by

$$m(C) = 1 - \frac{3}{2} \int_{D(C)} \left(\frac{\partial C}{\partial u} v + \frac{\partial C}{\partial v} u\right) du dv \quad (3-4)$$

where $D(C) = \{(u,v) \in \mathbf{I}^2 | C(u,v) < uv\}$.

Proposition 3.1 $0 \leq m(C) \leq 1$ for all $C \in \mathcal{C}$.

Proof: First, I am going to show $\int_{\mathbf{I}^2} \left(\frac{\partial C}{\partial u} v + \frac{\partial C}{\partial v} u\right) du dv = 2/3$ for all $C \in \mathcal{C}$.

A Measure for Positive Quadrant Dependence

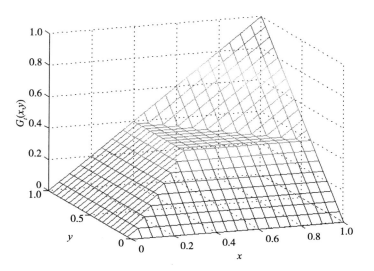

Figure 3–1　$C_t(u, v)$ with $t=0.4$

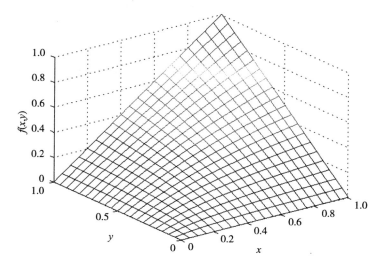

Figure 3–2　AMH copula with $\theta=0.8$

基于Copula的相关性测度
Measures of Association and Dependence Through Copulas

$$\int_0^1 \int_0^1 \left(\frac{\partial C}{\partial u}v + \frac{\partial C}{\partial v}u\right) dv du$$

$$= \int_0^1 \int_0^1 \frac{\partial C}{\partial u} v\, du\, dv + \int_0^1 \int_0^1 \frac{\partial C}{\partial v} u\, dv\, du$$

$$= \int_0^1 (C(1,v) - C(0,v))v\, dv + \int_0^1 (C(u,1) - C(u,0))u\, du$$

$$= \int_0^1 v^2 dv + \int_0^1 u^2 du$$

$$= 2/3$$

Now we have $\int_{D(C)} \left(\frac{\partial C}{\partial u}v + \frac{\partial C}{\partial v}u\right) du dv \leq 2/3$ for all C, since $D(C) \subseteq \leq I^2$ and $\frac{\partial C}{\partial u}, \frac{\partial C}{\partial v}, u, v$ are all non-negative. $\int_{D(C)} \left(\frac{\partial C}{\partial u}v + \frac{\partial C}{\partial v}u\right) du dv \geq 0$ is obvious, so we have the desired result.

Proposition 3.2 $m(C) = m(C^T)$, where $C^T(u, v) = C(v, u)$.

Proof: First, let us consider $m(C)$ on any rectangle $[m_1, m_2] \times [n_1, n_2]$ where $m_1, m_2, n_1, n_2 \in I$.

$$\int_{m_1}^{m_2} \int_{n_1}^{n_2} \left(\frac{\partial C}{\partial u}v + \frac{\partial C}{\partial v}u\right) dv du$$

$$= \int_{n_1}^{n_2} \int_{m_1}^{m_2} \frac{\partial C}{\partial u} v\, du\, dv + \int_{m_1}^{m_2} \int_{n_1}^{n_2} \frac{\partial C}{\partial v} u\, dv\, du$$

$$= \int_{n_1}^{n_2} (C(m_2,v) - C(m_1,v))v\, dv + \int_{m_1}^{m_2} (C(u,n_2) - C(u,n_1))u\, du$$

$$\int_{m_1}^{m_2} \int_{n_1}^{n_2} \left(\frac{\partial C^T}{\partial u}v + \frac{\partial C^T}{\partial v}u\right) dv du$$

$$= \int_{m_1}^{m_2} \int_{n_1}^{n_2} \frac{\partial C^T}{\partial u} v\, du\, dv + \int_{n_1}^{n_2} \int_{m_1}^{m_2} \frac{\partial C^T}{\partial v} u\, dv\, du$$

$$= \int_{m_1}^{m_2} (C^T(n_2,v) - C^T(n_1,v))v\, dv + \int_{n_1}^{n_2} (C^T(u,m_2) - C^T(u,m_1)) du du$$

$$= \int_{m_1}^{m_2} (C(v,n_2) - C(v,n_1))v\, dv + \int_{n_1}^{n_2} (C(m_2,u) - C(m_1,u))u\, du$$

So $m(C(u,v)) = m(C(v,u))$ on $[m_1, m_2] \times [n_1, n_2]$.

A Measure for Positive Quadrant Dependence

For any $S \in \mathcal{B}(\mathbf{I}^2)$, $m'(S) = \int_S \left(\frac{\partial C}{\partial u}v + \frac{\partial C}{\partial v}u\right) dudv$ defines a measure on the σ-algebra of $\mathbf{I}^2$. Since we have proved that $m(C) = m(C^T)$ for any rectangle $[m_1, m_2] \times [n_1, n_2]$, then according to the continuity of measure, $m(C) = m(C^T)$ on each $S \in \mathcal{B}(\mathbf{I}^2)$. Further, $m(C) = m(C^T)$ on $D(C)$.

Proposition 3.3 $m(C) = 1$ iff $C(u, v) \geq \Pi(u, v)$.

Proof: "$\Rightarrow$" $m(C) = 1$ implies $\int_{D(C)} \left(\frac{\partial C}{\partial u}v + \frac{\partial C}{\partial v}u\right) dudv = 0$. Claim, $\mu(D(C)) = 0$.

Suppose $\mu(D(C)) \neq 0$, then $\frac{\partial C}{\partial u}v + \frac{\partial C}{\partial v}u$ has to be 0, which implies $\frac{\partial C}{\partial u} = \frac{\partial C}{\partial v} = 0$ for all $(u,v) \in D(C)$. This is impossible (Here μ is the Lebesgue measure). Since C is continuous, there are no $(u,v) \in \mathbf{I}^2$ such that $C(u, v) < uv$, so $C \in PQD$.

"$\Leftarrow$" If $C \in PQD$, then $D(C) = \emptyset$. We have shown $\int_{\mathbf{I}^2} \left(\frac{\partial C}{\partial u}v + \frac{\partial C}{\partial v}u\right) dudv = 2/3$ in Proposition 3.1, so $m(C) = 1$.

Proposition 3.4 $m(C) = 0$ iff $C(u, v) < \Pi(u, v)$ for all $(u,v) \in \mathbf{I}^2$.

Proof: Denote $\overline{D(C)} = \mathbf{I}^2 / D(C)$.

"$\Rightarrow$" $m(C) = 0$ implies

$$\int_{D(C)} \left(\frac{\partial C}{\partial u}v + \frac{\partial C}{\partial v}u\right) dudv = \frac{2}{3}$$

further,

$$\int_{\overline{D(C)}} \left(\frac{\partial C}{\partial u}v + \frac{\partial C}{\partial v}u\right) dudv$$
$$= \int_{\mathbf{I}^2} \left(\frac{\partial C}{\partial u}v + \frac{\partial C}{\partial v}u\right) dudv - \int_{D(C)} \left(\frac{\partial C}{\partial u}v + \frac{\partial C}{\partial v}u\right) dudv$$
$$= 0$$

Since $\frac{\partial C}{\partial u}, \frac{\partial C}{\partial v}$ will not be 0 for all $(u,v) \in \overline{D(C)}$, then $\mu(\overline{D(C)}) = 0$, thus $\mu(D(C)) = 1$. Then $C(u,v)$ is continuous implies $D(C)$ is $\mathbf{I}^2$.

"$\Leftarrow$" If $C(u,v) < \Pi(u,v)$ for all $(u,v) \in \mathbf{I}^2$, then $D(C) = \mathbf{I}^2$. Since we have shown $\int_{\mathbf{I}^2} \left(\frac{\partial C}{\partial u}v + \frac{\partial C}{\partial v}u\right) dudv = 2/3$ in Proposition 3.1, then $m(C) = 0$.

基于Copula的相关性测度
Measures of Association and Dependence Through Copulas

Example 3.3 For Ali-Mikhail-Haq Copulas with $\theta > 0$, they are PQD. By applying (3-4), we get $m(C) = 1$.

Example 3.4 For the copula in Example 3.1, using the following R code, we will find $m(C) = 0.297$ for $t = 0.2$ and $m(C) = 0.944$ for $t = 0.8$.

④
Measures for Discrete MCD and Functional Dependence

In this section, we will define a measure of discrete MCD through conditional distribution and subcopula. Let C denote the set of discrete 2-subcopulas. Let $X \in S_1$, $S_1 = \{x_1, x_2, \cdots, x_{I+1}\}$ and $Y \in S_2$, $S_2 = \{y_1, y_2, \cdots, y_{J+1}\}$ ($I, J \in \{1, 2, \cdots, \infty\}$) be random variables with marginal distributions F_X, G_Y and joint distribution $H_{X,Y}$. Throughout this paper, we denote $u_i = F_X(x_i)$, $v_j = G_Y(y_j)$ and $u_0 = 0 = v_0$. For simplicity, we assume $P(X=x_i) \neq 0$, $i \in \{1, \cdots, I+1\}$ and $P(Y=y_j) \neq 0$, $j \in \{1, \cdots, J+1\}$. Consequently, $u_i \neq u_{i+1}$ and $v_j \neq v_{j+1}$ for all i and j. Denote the subcopula associated with the joint distribution of X and Y by C, then $C(F_X(x_i), G_Y(y_j)) = H_{X,Y}(x_i, y_j)$. The following lemma will be used in the proofs of our main results.

Lemma 4.1 Let $C \in \mathcal{C}$ be a subcopula, then
$$|C(u_j, v_j) - C(u_i, v_i)| \leq |u_j - u_i| + |v_j - v_i|$$
for any (u_i, v_i), (u_j, v_j) with $i < j$ in $D(C)$.

In particular
$$\frac{C(u_j, v_i) - C(u_i, v_i)}{u_j - u_i} \leq 1$$
provided $v_i = v_j$. Similarly,
$$\frac{C(u_i, v_j) - C(u_i, v_i)}{v_j - v_i} \leq 1$$
provided $u_i = u_j$.

基于Copula的相关性测度
Measures of Association and Dependence Through Copulas

4.1 The measure of MCD through conditional distributions

In Siburg and Stoimenov's paper, their measure is the standardized distance, which is induced by Sobolev norm, from the copula associated with X and Y to the independent copula. To define a measure of discrete MCD, we need a similar distance for discrete random variables.

Definition 4.1 For discrete random variables X and Y and constant $t \in I$, the expectation of weighted distance between the conditional distribution of Y given X and marginal distribution of Y is given by

$$\omega^2(Y \mid X) = t \sum_{i=0}^{I} \sum_{j=0}^{J} [P(Y \leq y_j \mid X = x_{i+1}) - G(y_j)]^2 f(x_{i+1}) g(y_{j+1}) +$$

$$(1-t) \sum_{i=0}^{I} \sum_{j=0}^{J} [P(Y \leq y_{j+1} \mid X = x_{i+1}) - G(y_{j+1})]^2 f(x_{i+1}) g(y_{j+1})$$

where $f(x_i) = P(X=x_i)$ and $g(y_j) = P(Y=y_j)$ are probability mass function (p.m.f.) of X and Y, respectively. Similarly, we can define $\omega^2(X \mid Y)$ as the expectation of weighted distance between the conditional distribution of X given Y and marginal distribution of X.

If we replace $G(y)$ in $\omega^2(Y \mid X)$ by $P(Y_I \leq y \mid X_I = x)$, where X_I and Y_I are independent variables with the same marginal distributions as X and Y respectively, then $\omega^2(Y \mid X)$ can be interpreted as a expected distance from the current conditional distribution function to the conditional distribution function when X and Y are independent. An interesting question is under what conditions ω^2 reaches its maximum value. We first introduce a notation.

Definition 4.2 Let X and Y be discrete random variables. The maximum of ω^2 is defined by

$$\omega^2_{\max}(Y \mid X) = \max \{\omega^2(Y \mid X) \mid H_{(X,Y)} \in \mathcal{D}_{(F,G)}\}$$

Measures for Discrete MCD and Functional Dependence

where $D_{(F,G)}$ is the class of joint distributions with marginals F and G respectively. Similarly, we can define $\omega_{\max}^2 (X \mid Y)$.

The following result gives the maximum values of $\omega^2 (Y \mid X)$ and $\omega^2 (X \mid Y)$, and the proof will be shown later in Theorem 4.1.

Proposition 4.1 For random variables X, Y with joint $H_{(X,Y)} \in D_{(F,G)}$, and marginals F, G respectively. we have

$$\omega_{\max}^2(Y \mid X) = \sum_{j=0}^{J} \{t[G(y_j) - (G(y_j))^2] + (1-t)[G(y_{j+1}) - (G(y_{j+1}))^2]\} g(y_{j+1})$$

and

$$\omega_{\max}^2(X \mid Y) = \sum_{i=0}^{I} \{t[F(x_i) - (F(x_i))^2] + (1-t)[F(x_{i+1}) - (F(x_{i+1}))^2]\} f(x_{i+1})$$

where $t \in I$ is a constant.

With ω^2 given above, we can define a measure of MCD as follows:

Definition 4.3 For any joint distribution $H_{(X,Y)} \in D_{(F,G)}$, the proper measure of MCD is given by

$$\mu_t(X, Y) = \left(\frac{\omega^2(Y \mid X) + \omega^2(X \mid Y)}{\omega_{\max}^2(Y \mid X) + \omega_{\max}^2(X \mid Y)} \right)^{\frac{1}{2}} \qquad (4-1)$$

where ω^2's are given in Definition 4.1.

Proposition 4.2 For any discrete random variables X and Y, the measure of discrete MCD μ_t has the following properties:

a. $\mu_t (X, Y) = \mu_t (Y, X)$;
b. $0 \leq \mu_t (X, Y) \leq 1$;
c. $\mu_t (X, Y) = 0$ if and only if X and Y are independent;
d. $\mu_t (X, Y) = 1$ if and only if X and Y are MCD.

Proof: Note that statement a and b are obvious from the definition of $\mu_t (X, Y)$. Next, we will prove c. Assume that X and Y are independent. Then

$$P(Y \leq y_j \mid X = x_i) = G(y_j), \text{ and } P(X \leq x_i \mid Y = y_j) = F(X_i)$$

for all i and j. Then from the definition of ω^2's, we obtain $\omega^2 (Y \mid X) = \omega^2 (X \mid Y) = 0$, and hence $\mu_t (X, Y) = 0$. Assume that $\mu_t (X, Y) = 0$, then $\omega^2 (Y \mid X) = \omega^2 (X \mid Y) = 0$ as both $\omega^2 (Y \mid X)$ and $\omega^2 (X \mid Y)$ are nonnegative.

基于Copula的相关性测度
Measures of Association and Dependence Through Copulas

Since every term in both $\omega^2 (Y|X)$ and $\omega^2 (X|Y)$ is nonnegative, we obtain $P(Y \leqslant y_j | X=x_i) - G(y_j) = 0$ and $P(X \leqslant x_i | Y=y_j) - F(X_i) = 0$ for all i and j. Therefore X and Y are independent.

In order to prove statement d, we need the following results:

Lemma 4.2 Let $X \in S_1$ and $Y \in S_2$ be two discrete random variables and $\mathbb{1}_A$ is the indicator function of set A, given by

$$\mathbb{1}_A = \begin{cases} 1 & \text{if } x \in A \\ 0 & \text{if } x \notin A \end{cases}$$

then the following two conditions are equivalent to each other:

a. There exists a function φ such that $Y = \varphi(X)$;

b. $E(\mathbb{1}_{Y \leqslant y_j} | X=x_i) = 0$ or 1 for all $(x_i, y_j) \in S_1 \times S_2$.

Similarly, we can conclude that the following two conditions are also equivalent to each other:

a. There exists a function ψ such that $X = \psi(Y)$;

b. $E(\mathbb{1}_{X \leqslant x_i} | Y=y_j) = 0$ or 1 for all $(x_i, y_j) \in S_1 \times S_2$.

Proof: Note that
$$E(I_{Y \leqslant y_j} | X = x_i) = 0 \quad \text{or} \quad 1$$

iff
$$P(Y \leqslant y_j | X = x_i) = 0 \quad \text{or} \quad 1$$

iff
$$\sum_{t \leqslant y_j} P(Y = t | X = x_i) = 0 \quad \text{or} \quad 1$$

iff
$$\sum_{t \leqslant y_j} P_{X,Y}(x_i, t) = 0 \quad \text{or} \quad P_X(x_i)$$

iff there exists a $y_0(x_i)$ such that
$$P_{X,Y}(x_i, y_j) = \begin{cases} P_X(x_i) & \text{if } y = y_0(x_i) \\ 0 & \text{if } y \neq y_0(x_i) \end{cases}$$

Therefore, the map $f(x_i) = y_0(x_i)$ is the desired function.

Measures for Discrete MCD and Functional Dependence

Notice that

$$P(Y \leq y_j \mid X = x_i) = \frac{P(X = x_i, Y \leq y_j)}{P_X(x_i)}$$

$$= \frac{P(X \leq x_i, Y \leq y_j) - P(X \leq x_{i-1}, Y \leq y_j)}{F_X(x_i) - F_X(x_{i-1})} \quad (4\text{-}2)$$

$$= \frac{C(u_i, v_j) - C(u_{i-1}, v_j)}{u_i - u_{i-1}}$$

for all $i = 1, \cdots, I+1$ and $j = 1, \cdots, J+1$.

Proposition 4.3 can be rephrased as:

There exists a Borel function f such that $Y = f(X)$ if and only if $\dfrac{C(u_i, v_j) - C(u_{i-1}, v_j)}{u_i - u_{i-1}}$ is 0 or 1. There exists a Borel function g such that $X = g(Y)$ if and only if $\dfrac{C(u_i, v_j) - C(u_i, v_{j-1})}{v_j - v_{j-1}}$ is 0 or 1.

Denote $P_{y_j \mid x_i} = P(Y \leq y_j \mid X = x_i)$.

Lemma 4.3 X and Y are discrete random variables with probability mass function f and g respectively, and $k, j \in J$, then

$$\sum_{i=0}^{I}(P_{y_k \mid x_{i+1}} - G(y_k))^2 f(x_{i+1}) g(y_{j+1}) = \sum_{i=0}^{I}(P^2_{y_k \mid x_{i+1}} - G(y_k)^2) f(x_{i+1}) g(y_{j+1})$$

Theorem 4.1 Let X and Y be discrete random variables with p.m.f. f and g, respectively. Then X and Y are MCD if and only if $\omega^2(Y \mid X) = \omega^2_{\max}(Y \mid X)$ and $\omega^2(X \mid Y) = \omega^2_{\max}(X \mid Y)$.

Proof: Note that for any X and Y, which are discrete random variables with p.m.f. f and g respectively, and for any $(x, y) \in S_1 \times S_2$, the following equality holds:

$$\sum_x [P(Y \leq y \mid X = x) - G(y)]^2 f(x) = \sum_x [P(Y \leq y \mid X = x)^2 - (G(y))^2] f(x)$$

Since $P(Y \leq y_j \mid X = x_{i+1}) \geq (P(Y \leq y_j \mid X = x_{i+1}))^2$, the first term of $\omega^2(Y \mid X)$ in (4.1) can be reduced to

基于Copula的相关性测度
Measures of Association and Dependence Through Copulas

$$\sum_{i=0}^{I}\sum_{j=0}^{J}[P(Y \leq y_j \mid X = x_{i+1}) - G(y_i)]^2 f(x_{i+1})g(y_{j+1})$$

$$= \sum_{i=0}^{I}\sum_{j=0}^{J}\{[P(Y \leq y_j \mid X = x_{i+1})]^2 - [G(y_i)]^2\}f(x_{i+1})g(y_{j+1})$$

$$\leq \sum_{i=0}^{I}\sum_{j=0}^{J}\{P(Y \leq y_j \mid X = x_{i+1}) - [G(y_i)]^2\}f(x_{i+1})g(y_{j+1})$$

$$= \sum_{i=0}^{I}\sum_{j=0}^{J} G(y_i)[1 - G(y_i)]f(x_{i+1})g(y_{j+1})$$

with equality holding if and only if $P(Y \leq y_j \mid X = x_{i+1})$ is either 0 or 1. Similarly, the second term of $\omega^2(Y \mid X)$ in (4.1) can be reduced to

$$\sum_{i=0}^{I}\sum_{j=0}^{J}[P(Y \leq y_{j+1} \mid X = x_{i+1}) - G(y_{i+1})]^2 f(x_{i+1})g(y_{j+1})$$

$$\leq \sum_{i=0}^{I}\sum_{j=0}^{J} G(y_{j+1})[1 - G(y_{j+1})]f(x_{i+1})g(y_{j+1})$$

with equality holding if and only if $P(Y \leq y_{j+1} \mid X = x_{i+1})$ is either 0 or 1. Then, $\omega^2(Y \mid X) = \omega_{\max}^2(Y \mid X)$ if and only if there exists a function φ such that $Y = \varphi(X)$. Using the same method, one can also show that $\omega^2(X \mid Y) = \omega_{\max}^2(X \mid Y)$ if and only if there exists a function ψ such that $Y = \psi(X)$. Combining those two together, we have X and Y are MCD if and only if $\omega^2(Y \mid X) = \omega_{\max}^2(Y \mid X)$ and $\omega^2(X \mid Y) = \omega_{\max}^2(X \mid Y)$.

Now Proposition 4.2 (d) follows

Remark. Proposition 4.1 is the consequence of Theorem 4.1. Indeed, we have

$$\omega^2(Y \mid X) \leq \sum_{j=0}^{J}\{t[G(y_j) - (G(y_j))^2] + (1-t)[G(y_{j+1}) - (G(y_{j+1}))^2]\}g(y_{j+1})$$

$$= \omega_{\max}^2(Y \mid X)$$

4 Measures for Discrete MCD and Functional Dependence

4.2 The measure of MCD through a subcopula

Although it is easy to understand the MCD measure $\mu_t(X, Y)$ as a proportion of expected weighted distance, its calculation is not easy to carry out. Note that there is a unique relationship between the joint distribution of (X, Y), and the subcopula, and it is easy to calculate the dependence relationship between X and Y through a subcopula. Therefore, it is necessary to define measures of MCD through subcopulas.

Notice that the conditional probabilities in Definition 4.3 can be replaced with subcopulas. For simplicity, we first introduce some notations, then the measure in subcopula form will be given as follows.

Definition 4.4 Let $C \in \mathcal{C}$ be a subcopula. The discrete norm of C is defined by

$$\|C\|_t^2 = \sum_{i=0}^{I}\sum_{j=0}^{J}\left\{(tC_{\Delta i,j}^2 + (1-t)C_{\Delta i,j+1}^2)\frac{\Delta v_j}{\Delta u_i} + (tC_{i,\Delta j}^2 + (1-t)C_{i+1,\Delta j}^2)\frac{\Delta u_i}{\Delta v_j}\right\}$$

$$= \sum_{i=0}^{I}\sum_{j=0}^{J}\left[t\left(\frac{C_{\Delta i,j}}{\Delta u_i}\right)^2 + (1-t)\left(\frac{C_{\Delta i,j+1}}{\Delta u_i}\right)^2\right]\Delta v_j \Delta u_i +$$

$$\sum_{i=0}^{I}\sum_{j=0}^{J}\left[t\left(\frac{C_{i,\Delta j}}{\Delta v_j}\right)^2 + (1-t)\left(\frac{C_{i+1,\Delta j}}{\Delta v_j}\right)^2\right]\Delta v_j \Delta u_i$$

where

$$C_{\Delta i,j} = C(u_{i+1}, v_j) - C(u_i, v_j)$$
$$C_{i,\Delta j} = C(u_i, v_{j+1}) - C(u_i, v_j)$$
$$\Delta u_i = u_{i+1} - u_i$$
$$\Delta v_j = v_{j+1} - v_j$$
$$t \in I$$

Theorem 4.2 Let C be a subcopula and $\Pi(u_i, v_j) = u_i v_j$ be the independent subcopula. Then, with notations given in Definition 4.4, $C(u_i, v_j) = \Pi(u_i, v_j)$ for all $i \in I$ and $j \in J$ if and only if

$$\|C\|_t^2 = \sum_{i=0}^{I}(tu_i^2 + (1-t)u_{i+1}^2)\Delta u_i + \sum_{j=0}^{J}(tv_j^2 + (1-t)v_{j+1}^2)\Delta v_j$$

基于Copula的相关性测度
Measures of Association and Dependence Through Copulas

Proof:

$$\|C-\Pi\|_t^2 = \sum_{i=0}^{I}\sum_{j=0}^{J} t[(C(u_{i+1},v_j)-u_{i+1}v_j)-(C(u_i,v_j)-u_iv_j)]^2 \frac{\Delta v_j}{\Delta u_i} +$$

$$\sum_{i=0}^{I}\sum_{j=0}^{J}(1-t)[(C(u_{i+1},v_{j+1})-u_{i+1}v_{j+1})-(C(u_i,v_{j+1})-u_iv_{j+1})]^2 \frac{\Delta v_j}{\Delta u_i} +$$

$$\sum_{i=0}^{I}\sum_{j=0}^{J} t[(C(u_i,v_{j+1})-u_iv_{j+1})-(C(u_i,v_j)-u_iv_j)]^2 \frac{\Delta u_i}{\Delta v_j} +$$

$$\sum_{i=0}^{I}\sum_{j=0}^{J}(1-t)[(C(u_{i+1},v_{j+1})-u_{i+1}v_{j+1})-(C(u_{i+1},v_j)-u_{i+1}v_j)]^2 \frac{\Delta u_i}{\Delta v_j}$$

$$= a_1 + a_2 + a_3 + a_4$$

where

$$a_1 = \sum_{i=0}^{I}\sum_{j=0}^{J} t[C_{\Delta i,j}-(u_{i+1}v_j-u_iv_j)]^2 \frac{\Delta v_j}{\Delta u_i}$$

$$= \sum_{i=0}^{I}\sum_{j=0}^{J} tC_{\Delta i,j}^2 \frac{\Delta v_j}{\Delta u_i} - \sum_{i=0}^{I}\sum_{j=0}^{J} 2tC_{\Delta i,j} v_j \Delta v_j + \sum_{i=0}^{I}\sum_{j=0}^{J} tv_j^2 \Delta u_i \Delta v_j$$

$$= \sum_{i=0}^{I}\sum_{j=0}^{J} tC_{\Delta i,j}^2 \frac{\Delta v_j}{\Delta u_i} - 2\sum_{j=0}^{J}\left\{\sum_{i=0}^{I} C_{\Delta i,j}\right\} tv_j \Delta v_j + \sum_{j=0}^{J} tv_j^2 \left\{\sum_{i=0}^{I} \Delta u_i\right\} \Delta v_j$$

$$= \sum_{i=0}^{I}\sum_{j=0}^{J} tC_{\Delta i,j}^2 \frac{\Delta v_j}{\Delta u_i} - \sum_{j=0}^{J} tv_j^2 \Delta v_j$$

Similarly, we can obtain

$$a_2 = \sum_{i=0}^{I}\sum_{j=0}^{J}(1-t)C_{\Delta i,j+1}^2 \frac{\Delta v_j}{\Delta u_i} - \sum_{i=0}^{I}(1-t)v_{j+1}^2 \Delta v_j$$

$$a_3 = \sum_{i=0}^{I}\sum_{j=0}^{J} tC_{i,\Delta j}^2 \frac{\Delta u_i}{\Delta v_j} - \sum_{i=0}^{I} tu_i^2 \Delta u_i$$

$$a_4 = \sum_{i=0}^{I}\sum_{j=0}^{J}(1-t)C_{i+1,\Delta j}^2 \frac{\Delta u_i}{\Delta v_j} - \sum_{i=0}^{I}(1-t)u_{i+1}^2 \Delta u_i$$

Combining a_1, a_2, a_3, and a_4 above, we obtain

$$\|C-\Pi\|_t^2 = \|C\|_t^2 - \sum_{j=0}^{J}(tv_j^2+(1-t)v_{j+1}^2)\Delta v_j - \sum_{i=0}^{I}(tu_i^2+(1-t)u_{i+1}^2)\Delta u_i$$

· 40 ·

Measures for Discrete MCD and Functional Dependence

Now it is clear that $C = \Pi$ implies

$$\|C\|_t^2 = \sum_{j=0}^{J}(tv_j^2 + (1-t)v_{j+1}^2)\,\Delta v_j + \sum_{i=0}^{I}(tu_i^2 + (1-t)u_{i+1}^2)\,\Delta u_i$$

On the other hand, if

$$\|C\|_t^2 = \sum_{j=0}^{J}(tv_j^2 + (1-t)v_{j+1}^2)\,\Delta v_j + \sum_{i=0}^{I}(tu_i^2 + (1-t)u_{i+1}^2)\,\Delta u_i$$

we must have

$$C(u_i, v_{j+1}) - u_i v_{j+1} = C(u_i, v_j) - u_i v_j$$

$$= C(u_{i+1}, v_j) - u_{i+1} v_j = C(u_{i+1}, v_{j+1}) - u_{i+1} v_{j+1}$$

for all i and j. Therefore $C(u_i, v_j) = u_i v_j$ for all i and j.

From Theorem 4.2, it is easy to prove the following result.

Corollary 4.1 For any $C \in \mathcal{C}$, the following hold:

$$\sum_{i=0}^{I}\sum_{j=0}^{J} C_{\Delta i, j}^2 \frac{\Delta v_j}{\Delta u_i} \geqslant \sum_{j=0}^{J} v_j^2 \,\Delta v_j$$

$$\sum_{i=0}^{I}\sum_{j=0}^{J} C_{\Delta i, j+1}^2 \frac{\Delta v_j}{\Delta u_i} \geqslant \sum_{j=0}^{J} v_{j+1}^2 \,\Delta v_j$$

$$\sum_{i=0}^{I}\sum_{j=0}^{J} C_{i, \Delta j}^2 \frac{\Delta u_i}{\Delta v_j} \geqslant \sum_{j=0}^{J} u_i^2 \,\Delta u_i$$

$$\sum_{i=0}^{I}\sum_{j=0}^{J} C_{i+1, \Delta j}^2 \frac{\Delta u_i}{\Delta v_j} \geqslant \sum_{j=0}^{J} u_{i+1}^2 \,\Delta u_i$$

Given the joint distribution $H_{X,Y}$ and its corresponding subcopula C, we say that C is MCD if and only if X and Y are MCD.

Theorem 4.3 For $C \in \mathcal{C}$, C is MCD if and only if

$$\|C\|_t^2 = \sum_{i=0}^{I}(tu_i + (1-t)u_{i+1})\,\Delta u_i + \sum_{j=0}^{J}(tv_j + (1-t)v_{j+1})\,\Delta v_j$$

Proof:

基于Copula的相关性测度
Measures of Association and Dependence Through Copulas

$$\|C\|_t^2 = \sum_{i=0}^{I}\sum_{j=0}^{J}\{c_{\Delta i,j} + c_{i,\Delta j}\}\, \Delta u_i\, \Delta v_j$$

$$\leqslant \sum_{i=0}^{I}\sum_{j=0}^{J}\{c_{\Delta i,j} + c_{i,\Delta j}\}\, \Delta u_i\, \Delta v_j$$

$$= \sum_{j=0}^{J}(tv_j + (1-t)v_{j+1})\, \Delta v_j + \sum_{i=0}^{I}(tu_i + (1-t)u_{i+1})\, \Delta u_i$$

with equality holds if and only if both $\frac{C_{i,\Delta j}}{\Delta v_j}$ and $\frac{C_{\Delta i,j}}{\Delta u_i}$ are 0 or 1 for all i and j, where
$c_{\Delta i,j} = t\left(\frac{C_{\Delta i,j}}{\Delta u_i}\right)^2 + (1-t)\left(\frac{C_{\Delta i,j+1}}{\Delta u_i}\right)^2$ and $c_{i,\Delta j} = t\left(\frac{C_{i,\Delta j}}{\Delta v_j}\right)^2 + (1-t)\left(\frac{C_{i+1,\Delta j}}{\Delta v_j}\right)^2$.

Corollary 4.2 Let X and Y be two discrete random variables and C be the subcopula associated with $H_{X,Y}$, then

$$\sum_{i=0}^{I}\sum_{j=0}^{J}\left(\frac{C(u_{i+1},v_j) - C(u_i,v_j)}{\Delta u_i}\right)^2 \Delta u_i\, \Delta v_j \leqslant \sum_{j=0}^{J} v_j\, \Delta v_j$$

and

$$\sum_{i=0}^{I}\sum_{j=0}^{J}\left(\frac{C(u_{i+1},v_{j+1}) - C(u_i,v_{j+1})}{\Delta u_i}\right)^2 \Delta u_i\, \Delta v_j \leqslant \sum_{j=0}^{J} v_{j+1}\, \Delta v_j$$

with equality holds if and only if there exists a Borel function f such that $Y = f(X)$.

$$\sum_{i=0}^{I}\sum_{j=0}^{J}\left(\frac{C(u_i,v_{j+1}) - C(u_i,v_j)}{\Delta v_j}\right)^2 \Delta u_i\, \Delta v_j \leqslant \sum_{i=0}^{I} u_i\, \Delta u_i$$

and

$$\sum_{i=0}^{I}\sum_{j=0}^{J}\left(\frac{C(u_{i+1},v_{j+1}) - C(u_{i+1},v_j)}{\Delta v_j}\right)^2 \Delta u_i\, \Delta v_j \leqslant \sum_{i=0}^{I} u_{i+1}\, \Delta u_i$$

with equality holds if and only if there exists a Borel function g such that $X = g(Y)$.

Summarizing Theorems 4.2, Theorems 4.3 and Corollaries 4.1, Corollaries 4.2, we obtain the range of $\|C\|_t^2$:

Theorem 4.4 For any $C(u_i, v_j) \in C$ and $t \in I$,

$$L_t \leqslant \|C\|_t^2 \leqslant U_t$$

where

4 Measures for Discrete MCD and Functional Dependence

$$L_t = L_t^{(1)} + L_t^{(2)} = \sum_{i=0}^{I}(tu_i^2 + (1-t)u_{i+1}^2)\,\Delta u_i + \sum_{j=0}^{J}(tv_j^2 + (1-t)v_{j+1}^2)\,\Delta v_j$$

and

$$U_t = U_t^{(1)} + U_t^{(2)} = \sum_{i=0}^{I}(tu_i + (1-t)u_{i+1})\,\Delta u_i + \sum_{j=0}^{J}(tv_j + (1-t)v_{j+1})\,\Delta v_j$$

Notice that the lower bound of $\|C\|_t^2$ depends on the marginal distributions of X and Y, in order to compare MCD of copulas with different marginal distributions, we need to introduce the following standardized form.

Definition 4.5 Given joint distribution of two discrete random variables X, Y associated with a subcopula C, the proper measure for MCD is defined by

$$\mu_t(X,Y) = \left(\frac{\|C\|_t^2 - L_t}{U_t - L_t}\right)^{\frac{1}{2}} \quad (4-3)$$

Note that $\mu_t(X,Y) \in I$, which measures mutual functional relationship between X and Y.

Remark. The MCD measure $\mu_t(X,Y)$ in (4-1) and the measure in (4-3) are equivalent:

$$\mu_t(X,Y) = \left(\frac{\omega^2(Y\mid X) + \omega^2(X\mid Y)}{\omega_{\max}^2(Y\mid X) + \omega_{\max}^2(X\mid Y)}\right)^{\frac{1}{2}} = \left(\frac{\|C\|_t^2 - L_t}{U_t - L_t}\right)^{\frac{1}{2}}$$

Indeed, note that $u_i = F(x_i)$ and $v_j = G(y_j)$ for all i and j. From Proposition 4.1, we have

$$\omega_{\max}^2(Y\mid X) = \sum_{j=0}^{J}\{t[v_j - v_j^2] + (1-t)[v_{j+1} - v_{j+1}^2]\}\,\Delta v_j$$

and

$$\omega_{\max}^2(X\mid Y) = \sum_{i=0}^{I}\{t[u_i - u_i^2] + (1-t)[u_{i+1} - u_{i+1}^2]\}\,\Delta u_i$$

So that $\omega_{\max}^2(Y\mid X) + \omega_{\max}^2(X\mid Y) = U_t \in L_t$. Similarly, we can obtain $\omega^2(Y\mid X) + \omega^2(X\mid Y) = \|C\|_t^2 \in L_t$.

Remark. The discrete norm of C given in Definition 4.4 is consistent with the continuous norm given in Siburg and Stoimenov. Indeed, if $\max\{\Delta u_i, \Delta v_j\} \to 0$, then

基于Copula的相关性测度
Measures of Association and Dependence Through Copulas

$$\|C\|_t^2 = \sum_{i=0}^{I}\sum_{j=0}^{J}\left(\frac{tC_{\Delta i,j}^2}{\Delta u_i^2} + \frac{(1-t)C_{\Delta i,j+1}^2}{\Delta u_i^2} + \frac{tC_{i,\Delta j}^2}{\Delta v_j^2} + \frac{(1-t)C_{i+1,\Delta j}^2}{\Delta v_j^2}\right)\Delta u_i\,\Delta v_j$$

$$\to \int_0^1\int_0^1\left[\left(\frac{\partial C}{\partial u}\right)^2 + \left(\frac{\partial C}{\partial v}\right)^2\right]du\,dv$$

Also we obtain $L_t \to 2/3$, and $U_t \to 1$ as $\max\{\Delta u_i, \Delta v_j\} \to 0$.

Lemma 4.4 For any discrete random variables X and Y.

$$\sum_{i=1}^{I+1}\sum_{j=1}^{J+1}(P(Y\leqslant y_j \mid X=x_i) - G(y_j))^2 f(x_i)g(y_j)$$

$$=\sum_{i=1}^{I+1}\sum_{j=1}^{J+1}(P(Y\leqslant y_j \mid X=x_i)^2 - G(y_j)^2)f(x_i)g(y_j)$$

Proof: From the fact that $\sum_x\sum_y P(Y\leqslant y \mid X=x)G(y)f(x)g(y) = \sum_y G(y)^2 g(y)$,

$$\sum_x\sum_y\left(P(Y\leqslant y \mid X=x) - G(y)\right)^2 f(x)g(y)$$

$$=\sum_x\sum_y\left(P(Y\leqslant y \mid X=x)^2 - 2P(Y\leqslant y \mid X=x)G(y) + G(y)^2\right)f(x)g(y)$$

$$=\sum_x\sum_y\left(P(Y\leqslant y \mid X=x)^2\right)f(x)g(y) -$$

$$2\sum_x\sum_y P(Y\leqslant y \mid X=x)G(y)f(x)g(y) + \sum_x\sum_y G(y)^2 f(x)g(y)$$

$$=\sum_x\sum_y\left(P(Y\leqslant y \mid X=x)^2\right)f(x)g(y) - 2\sum_y G(y)^2 g(y) + \sum_y G(y)^2 g(y)$$

$$=\sum_x\sum_y\left(P(Y\leqslant y \mid X=x)^2\right)f(x)g(y) - \sum_y G(y)^2 g(y)$$

$$=\sum_x\sum_y\left(P(Y\leqslant y \mid X=x)^2 - G(y)^2\right)f(x)g(y)$$

which completes the proof.

But in some situations, the functional relationship may not be mutual. For example, Y could be a function of X, but the reverse may not be true. So next, we will modify μ_t in order to obtain the measures for this type of functional relationships.

Definition 4.6 Given the joint distribution of two discrete random variables

Measures for Discrete MCD and Functional Dependence

X, Y associated with a subcopula C, two measures $\mu_t(Y \mid X)$ and $\mu_t(X \mid Y)$ for $Y=f(X)$ and $X=g(Y)$, respectively, are defined as follows:

$$\mu_t(Y \mid X) = \left(\frac{\sum_{i=0}^{I} \sum_{j=0}^{J} c_{\Delta i, j} \Delta u_i \Delta v_j - \sum_{j=0}^{J} (t v_j^2 + (1-t) v_{j+1}^2) \Delta v_j}{\sum_{j=0}^{J} (t v_j + (1-t) v_{j+1}) \Delta v_j - \sum_{j=0}^{J} (t v_j^2 + (1-t) v_{j+1}^2) \Delta v_j} \right)^{\frac{1}{2}}$$

and

$$\mu_t(X \mid Y) = \left(\frac{\sum_{i=0}^{I} \sum_{j=0}^{J} c_{i, \Delta j} \Delta u_i \Delta v_j - \sum_{i=0}^{I} (t u_i^2 + (1-t) u_{i+1}^2) \Delta u_i}{\sum_{i=0}^{I} (t u_i + (1-t) u_{i+1}) \Delta u_i - \sum_{i=0}^{I} (t u_i^2 + (1-t) u_{i+1}^2) \Delta u_i} \right)^{\frac{1}{2}}$$

where $c_{\Delta i,j} = t \left(\frac{C_{\Delta i,j}}{\Delta u_i} \right)^2 + (1-t) \left(\frac{C_{\Delta i,j+1}}{\Delta u_i} \right)^2$ and $c_{i, \Delta j} = t \left(\frac{C_{i, \Delta j}}{\Delta v_j} \right)^2 + (1-t) \left(\frac{C_{i+1, \Delta j}}{\Delta v_j} \right)^2$.

Our next result will show that μ_t, $\mu_t(Y \mid X)$, and $\mu_t(X \mid Y)$ are the proper measures.

Theorem 4.5 For any discrete random variables X and Y with a subcopula C, μ_t's defined above have the following properties:

a. $\mu_t(X, Y) = \mu_t(Y, X)$;
b. $0 \leq \mu_t(X, Y) \leq 1$;
c. $\mu_t(X, Y) = 0$ if and only if X and Y are independent;
d. $\mu_t(X, Y) = 1$ if and only if X and Y are MCD;
e. $\mu_t(Y \mid X) = 1$ if and only if $Y=f(X)$ for some Borel function f;
f. $\mu_t(X \mid Y) = 1$ if and only if $X=g(Y)$ for some Borel function g.

Proof: (a) is obvious by the definition of μ_t. (b), (c), and (d) are direct results of Theorem 4.2, Theorem 4.3, and Definition 4.5. Also (e) and (f) can be derived from Corollary 4.1 and Corollary 4.2. Note that t is a weight parameter of those measures. In general, we choose $t = \frac{1}{2}$. Thus

$$\sum_{i=0}^{I} \left(\frac{1}{2} u_i + \frac{1}{2} u_{i+1} \right) \Delta u_i = \frac{1}{2} = \sum_{j=0}^{J} \left(\frac{1}{2} v_j + \frac{1}{2} v_{j+1} \right) \Delta v_j$$

and

$$\mu_{\frac{1}{2}}(X, Y) = \left(\frac{\|C\|_{\frac{1}{2}}^2 - L_{\frac{1}{2}}}{1 - L_{\frac{1}{2}}} \right)^{\frac{1}{2}}$$

基于Copula的相关性测度
Measures of Association and Dependence Through Copulas

Example 4.1 Let X has the distribution as showed in Table 4-1 and $Y = X^2$. The joint distribution and the corresponding subcopula of X and Y are in Table 4-2 and Table 4-3 respectively. According to Definition 4.5, we have $\mu_t = \sqrt{\dfrac{32t+77}{32t+137}}$, $\mu_{\frac{1}{2}} = 0.78$, $\mu_{\frac{1}{2}}(Y|X) = 1$, and $\mu_{\frac{1}{2}}(X|Y) = \dfrac{\sqrt{21}}{9}$.

Table 4-1 Distribution of X

$X=x$	-2	-1	0	1	2
$P(X=x)$	2/8	1/8	2/8	1/8	2/8

Table 4-2 Joint Distribution of X and Y

$Y=y$ \ $X=x$	-2	-1	0	1	2
0	0	0	2/8	0	0
1	0	1/8	0	1/8	0
4	2/8	0	0	0	2/8

Example 4.2 Consider a sequence of independent and identical Bernoulli trial with success rate p. Let X be the number of success in the first trial, $Y=2^k$ where k is the number of trials until first success. Thus joint mass function of X and Y is given in Table 4-2 and the corresponding subcopula is in Table 4-2.

Table 4-3 Subcopula of X and Y

$V=v$ \ $U=u$	1/4	3/8	5/8	3/4	1
1/4	0	0	1/4	1/4	1/4
1/2	0	1/8	3/8	1/2	1/2
1	1/4	3/8	5/8	3/4	1

Measures for Discrete MCD and Functional Dependence

Table 4–4 Joint Distribution of X and Y

X=x \ Y=y	2	4	8	16	...
0	0	$p(1-p)$	$p(1-p)^2$	$p(1-p)^3$	...
1	p	0	0	0	...

From Table 4-2, we have

$$v_i = 1 - (1-p)^i \qquad \Delta v_i = p(1-p)^i \qquad \text{for all} \qquad i \geqslant 0,$$

$$u_0 = 0 \qquad u_1 = 1 - p \qquad u_2 = 1.$$

Table 4–5 Corresponding subcopula of X and Y

U=u \ V=v	p	$p+p(1-p)$	...	$1-(1-p)^{n+1}$	...
$1-p$	0	$p(1-p)$	...	$1-p-(1-p)^{n+1}$	...
1	p	$p+p(1-p)$	...	$1-(1-p)^{n+1}$	...

From the joint distribution, it is obvious that X is a function of Y. One can also verify this by calculating $\mu_t(X \mid Y)$.

$$\sum_{i \geqslant 0}\sum_{j \geqslant 0}\left[t\left(\frac{C_{i,\Delta j}}{\Delta v_j}\right)^2 + (1-t)\left(\frac{C_{i+1,\Delta j}}{\Delta v_j}\right)^2\right]\Delta u_i \Delta v_j$$

$$= \sum_{j \geqslant 0}\left[t\left(\frac{C_{0,\Delta j}}{\Delta v_j}\right)^2 + (1-t)\left(\frac{C_{1,\Delta j}}{\Delta v_j}\right)^2\right]\Delta u_0 \Delta v_j$$

$$= \sum_{j \geqslant 0}\left[t\left(\frac{C_{1,\Delta j}}{\Delta v_j}\right)^2 + (1-t)\left(\frac{C_{2,\Delta j}}{\Delta v_j}\right)^2\right]\Delta u_1 \Delta v_j$$

$$= (1-t)(1-p)^2 + (1-t)p^2 + p(1-p)$$

$$= (1-t)(1-p)^2 + (t(1-p) + (1-t))p$$

$$= \sum_{i=0}^{I}(tu_i + (1-t)u_{i+1})\Delta u_i$$

Therefore, $\mu_t(X \mid Y) = 1$, which implies that X is a function of Y. But Y is not a function of X, so next we calculate $\mu_t(Y \mid X)$. Note that the terms of μ_t

基于Copula的相关性测度
Measures of Association and Dependence Through Copulas

$(Y|X)$ in Definition 2.4 are simplified as

$$\sum_{i>0}\sum_{j>0}\left(\frac{C_{\Delta i,j}}{\Delta u_i}\right)^2 \Delta u_i \Delta v_j = \frac{(p-1)(p^3-2p^2+2)}{(p-2)(p^2-3p+3)}$$

$$\sum_{i>0}\sum_{j>0}\left(\frac{C_{\Delta i,j+1}}{\Delta u_i}\right)^2 \Delta u_i \Delta v_j = p - \frac{(p-1)^2(p^2-2p+2)}{(p-2)(p^2-3p+3)}$$

$$\sum_{j>0}(v_j^2)\Delta v_j = \frac{1}{p^2-3p+3}+\frac{2}{p-2}+1$$

$$\sum_{j>0}(v_{j+1}^2)\Delta v_j = \frac{p-2}{p^2-3p+3}-\frac{2}{p-2}$$

$$\sum_{j>0}(v_j)\Delta v_j = 1+\frac{1}{p-2}$$

$$\sum_{j>0}(v_{j+1})\Delta v_j = -\frac{1}{p-2}$$

Then plug them into $\mu_t(Y|X)$ in Definition 4.6, we have

$$\mu_t(Y|X) = \sqrt{p(2-p)}$$

Remark. This interesting example shows that the measure may not always depends on t;

Note that $E(Y)$ does not exist for $p \geqslant \frac{1}{2}$. This example also shows us that even though variable Y has no expectation, we can still calculate $\mu_t^{(1)}$ and $\mu_t(X|Y)$.

4.3 Comparison to Siburg and Stoimenov's measure of MCD

First we show that the Siburg and Stoimenov's measure of MCD may not fitted in computing discrete MCD.

4.3.1 Extension using E-process

E. de Amo et al. proposed the way to extend a subcopula to a copula, called

Measures for Discrete MCD and Functional Dependence

E-process. Moreover, this process provides the way to obtain the copula which is the upper bound of the set of copula extending given subcopula. We will use this process to explain why Siburg and Stoimenov's measure is not appropriate for discrete setting.

Assume that we consider the independent subcopula $P(u,v) = uv$ where $u,v \in \{0, \frac{1}{2}, 1\}$. It is not hard to know that the independent copula $\Pi(u,v) = uv$ where $u,v \in I$ is one of the extension of P since $\Pi(u,v) = P(u,v)$ for all $(u,v) \in \{0, \frac{1}{2}, 1\} \times \{0, \frac{1}{2}, 1\}$. We know that for the copula Π, Siburg and Stoimenov's measure, denoted by ω, always takes value 0, i.e., $\omega(\Pi) = 0$. This implies that $\omega(P) = 0$ if we extend P to Π. Next, we will use E-process to extend the independent subcopula P and then compute the value of the measure ω. By E-process, we can obtain the exact form of the copula that extends the independent subcopula P, denoted by P^*, and hence get the upper bound of the set of copula extending independent subcopula, denoted by UP^*, as follows:

$$P^*(u,v) = \begin{cases} \frac{1}{4} C_{00}(F_{00}(2u), G_{00}(2v)) & \text{if } (u,v) \in [0, \frac{1}{2}] \times [0, \frac{1}{2}] \\ \frac{1}{4} C_{10}(F_{10}(2u-1), G_{10}(2v)) + \frac{1}{4} G_{00}(2v) & \text{if } (u,v) \in [\frac{1}{2}, 1] \times [0, \frac{1}{2}] \\ \frac{1}{4} C_{01}(F_{01}(2u), G_{01}(2v-1)) + \frac{1}{4} F_{00}(2u) & \text{if } (u,v) \in [0, \frac{1}{2}] \times [\frac{1}{2}, 1] \\ \frac{1}{4} + \frac{1}{4} C_{11}(F_{11}(2u-1), G_{11}(2v-1)) \\ \quad + \frac{1}{4} F_{10}(2u-1) + \frac{1}{4} G_{01}(2v-1) & \text{if } (u,v) \in [\frac{1}{2}, 1] \times [\frac{1}{2}, 1] \end{cases}$$

where $C_{ij} \in C$ and F_{ij}, G_{ij} are distribution functions satisfying E-process for any $t, j \in \{0, 1\}$, and

$$UP^*(u,v) = \begin{cases} \frac{1}{4} \min\{F_{00}(2u), G_{00}(2v)\} & \text{if } (u,v) \in [0, \frac{1}{2}] \times [0, \frac{1}{2}] \\ \frac{1}{4} \min\{F_{10}(2u-1), G_{10}(2v)\} + \frac{1}{4} G_{00}(2v) & \text{if } (u,v) \in [\frac{1}{2}, 1] \times [0, \frac{1}{2}] \\ \frac{1}{4} \min\{F_{01}(2u), G_{01}(2v-1)\} + \frac{1}{4} F_{00}(2u) & \text{if } (u,v) \in [0, \frac{1}{2}] \times [\frac{1}{2}, 1] \\ \frac{1}{4} + \frac{1}{4} \min\{F_{11}(2u-1), G_{11}(2v-1)\} \\ \quad + \frac{1}{4} F_{10}(2u-1) + \frac{1}{4} G_{01}(2v-1) & \text{if } (u,v) \in [\frac{1}{2}, 1] \times [\frac{1}{2}, 1] \end{cases}$$

where

基于Copula的相关性测度
Measures of Association and Dependence Through Copulas

$$F_{00}(2u) = \begin{cases} 0 & u = 0 \\ 4u & 0 \leqslant u \leqslant \frac{1}{4} \\ 1 & \frac{1}{4} \leqslant u \leqslant \frac{1}{2} \end{cases}, \quad G_{00}(2v) = \begin{cases} 0 & v = 0 \\ 4v & 0 \leqslant v \leqslant \frac{1}{4} \\ 1 & \frac{1}{4} \leqslant v \leqslant \frac{1}{2} \end{cases},$$

$$F_{10}(2u-1) = \begin{cases} 0 & u = \frac{1}{2} \\ 4u - 2 & \frac{1}{2} \leqslant u \leqslant \frac{3}{4} \\ 1 & \frac{3}{4} \leqslant u \leqslant 1 \end{cases}, \quad G_{10}(2v) = \begin{cases} 0 & 0 \leqslant v \leqslant \frac{1}{4} \\ 4v - 1 & \frac{1}{4} \leqslant v \leqslant \frac{1}{2} \\ 1 & v = \frac{1}{2} \end{cases},$$

$$F_{01}(2u) = \begin{cases} 0 & 0 \leqslant u \leqslant \frac{1}{4} \\ 4u - 1 & \frac{1}{4} \leqslant v \leqslant \frac{1}{2} \\ 1 & u = \frac{1}{2} \end{cases}, \quad G_{01}(2v-1) = \begin{cases} 0 & v = \frac{1}{2} \\ 4v - 2 & \frac{1}{2} \leqslant v \leqslant \frac{3}{4} \\ 1 & \frac{3}{4} \leqslant v \leqslant 1 \end{cases},$$

$$F_{11}(2u-1) = \begin{cases} 0 & \frac{1}{2} \leqslant u \leqslant \frac{3}{4} \\ 4u - 3 & \frac{3}{4} \leqslant u \leqslant 1 \\ 1 & u = 1 \end{cases}, \quad G_{11}(2v-1) = \begin{cases} 0 & \frac{1}{2} \leqslant v \leqslant \frac{3}{4} \\ 4v - 3 & \frac{3}{4} \leqslant v \leqslant 1 \\ 1 & v = 1 \end{cases}.$$

Lemma 4.5 Let UP^* be the upper bound of the set of copula extending independent subcopula defined above. Then

$$\omega(UP^*) = 1$$

Proof: The modified Sobolev norm of UP^* is defined by

$$|UP^*|^2 = \int_0^1 \int_0^1 \left[\left(\frac{\partial UP^*}{\partial u} \right)^2 + \left(\frac{\partial UP^*}{\partial v} \right)^2 \right] dudv$$

Now, we distinguish the calculation into four cases.

Case 1. If $(u, v) \in [0, \frac{1}{2}] \times [0, \frac{1}{2}]$, then

$$\int_0^{\frac{1}{2}} \int_0^{\frac{1}{2}} \left(\frac{\partial UP^*}{\partial u} \right)^2 dudv = \frac{3}{32} \quad \text{and} \quad \int_0^{\frac{1}{2}} \int_0^{\frac{1}{2}} \left(\frac{\partial UP^*}{\partial v} \right)^2 dudv = \frac{3}{32}$$

Case 2. If $(u, v) \in [\frac{1}{2}, 1] \times [0, \frac{1}{2}]$, then

$$\int_0^{\frac{1}{2}} \int_{\frac{1}{2}}^1 \left(\frac{\partial UP^*}{\partial u} \right)^2 dudv = \frac{1}{32} \quad \text{and} \quad \int_0^{\frac{1}{2}} \int_{\frac{1}{2}}^1 \left(\frac{\partial UP^*}{\partial v} \right)^2 dudv = \frac{7}{32}$$

Case 3. If $(u, v) \in [0, \frac{1}{2}] \times [\frac{1}{2}, 1]$, then

$$\int_{\frac{1}{2}}^1 \int_0^{\frac{1}{2}} \left(\frac{\partial UP^*}{\partial u} \right)^2 dudv = \frac{7}{32} \quad \text{and} \quad \int_{\frac{1}{2}}^1 \int_0^{\frac{1}{2}} \left(\frac{\partial UP^*}{\partial v} \right)^2 dudv = \frac{1}{32}$$

Measures for Discrete MCD and Functional Dependence

Case 4. If $(u, v) \in [\frac{1}{2}, 1] \times [\frac{1}{2}, 1]$, then

$$\int_{\frac{1}{2}}^{1}\int_{\frac{1}{2}}^{1} \left(\frac{\partial UP^*}{\partial u}\right)^2 dudv = \frac{5}{32} \text{ and } \int_{\frac{1}{2}}^{1}\int_{\frac{1}{2}}^{1} \left(\frac{\partial UP^*}{\partial v}\right)^2 dudv = \frac{5}{32}$$

Hence, the value of modified Sobolev norm of the copula UP^* is

$$|UP^*|^2 = \left(\frac{3}{32} + \frac{1}{32} + \frac{7}{32} + \frac{5}{32}\right) + \left(\frac{3}{32} + \frac{7}{32} + \frac{1}{32} + \frac{5}{32}\right) = 1$$

Therefore, Siburg and Stoimenov's measure of UP^* is

$$\omega(UP^*) = (3|UP^*|^2 - 2)^{\frac{1}{2}} = 1.$$

Since we know that UP^* is one of the extension of P, from the previous lemma, we have $\omega(P) = 1$ if we extend P to UP^*. Furthermore, we can obtain the following consequence:

Theorem 4.6 Let P be the independent subcopula defined by $P(u, v) = uv$ where $u, v \in \{0, \frac{1}{2}, 1\}$. Then

$$\{\omega(C) \mid C \text{ is the extension of } P\} = I$$

Proof: It is clear that $\{\omega(C) \mid C \text{ is the extension of } P\} \subseteq I$. Let $t \in I$ and define $C_t = t\Pi + (1-t)UP^*$ where Π is the independent copula and UP^* is the upper bound of the set of copula extending independent subcopula. Then we see that $\omega(C_1) = 0$ and $\omega(C_0) = 1$. Since C is the linear combination of two copulas, C_t is a copula. Moreover, we can see that, for any $t \in I$, C_t is exactly the extension of P. This implies that $\{\omega(C) \mid C \text{ is the extension of } P\} \supseteq \{\omega(C_t) \mid t \in I\} = I.$

As a result, there are a few drawbacks of the measure ω. First, the value of ω is varied. We can see that the measure ω of the independent subcopula P takes value 0 and 1 if we extend P to the copula Π and UP^*, respectively. Furthermore, from the previous theorem, the measure ω can take value between 0 and 1 for the extension of P; this value depends on the extension of the subcopula. Second, we might expect that the measure ω of P should take value 0 since discrete random variables associated to P are independent. But, indeed, $\omega(P) = 1$ if

基于Copula的相关性测度
Measures of Association and Dependence Through Copulas

we extend P to UP^*. This shows that the measure ω can not measure MCD correctly.

4.3.2 Bilinear extension

Bilinear extension is the most popular way to construct a copula from a subcopula. It is defined as follows:

Let (u, v) be any point in I^2, let u_1 and u_2 be, respectively, the greatest and least elements of D_1 that satisfy $u_1 \leq u \leq u_2$; and let v_1 and v_2 be respectively, the greatest and least elements of D_2 that satisfy $v_1 \leq v \leq v_2$. Now define

$$C_{ex}(u,v) = (1-\lambda_1)(1-\mu_1)C(u_1,v_1) + (1-\lambda_1)\mu_1 C(u_1,v_2) + \\ \lambda_1(1-\mu_1)C(u_2,v_1) + \lambda_1\mu_1 C(u_2,v_2) \quad (4-4)$$

where $\lambda_1 = (u-u_1)/(u_2-u_1)$ and $\mu_1 = (v-v_1)/(v_2-v_1)$.

Since Siburg and Stoimenov already defined Sobolev norm for a copula, applying their definition to a bilinear extension of a subcopula may give another way to define a measure for discrete variables. But it is not feasible. We know the norm of a copula is in $[2/3, 1]$ and it reaches $2/3$ when the copula is a independent copula, it reaches 1 when the copula is MCD. But from the next theorem, we will see that the Sobolev norm of a bilinear extension of a subcopula is always strictly less than 1. In other words, subcopula is MCD doesn't implies bilinear extension is also MCD, which is consistent with the fact that bilinear extension will never reaches the Fréchet–Hoeffding bounds.

Theorem 4.7 Let $C \in \mathcal{C}$, C_{ex} is the bilinear extension of C as defined above. Then we have

$$|C_{ex}|^2 \leq 1$$

Proof: The Sobolev norm of a copula given by Siburg and Stoimenov is

$$|C_{ex}|^2 = \int_0^1 \int_0^1 \left[\left(\frac{\partial C_{ex}}{\partial u}\right)^2 + \left(\frac{\partial C_{ex}}{\partial v}\right)^2 \right] dudv$$

Measures for Discrete MCD and Functional Dependence

$$\int_0^1 \int_0^1 \left(\frac{\partial C_{ex}}{\partial u}\right)^2 dudv = \sum_{i=0}^{I} \sum_{j=0}^{J} \int_{v_j}^{v_{j+1}} \int_{u_i}^{u_{i+1}} \left(\frac{\partial C_{ex}}{\partial u}\right)^2 dudv$$

$$= \sum_{i=0}^{I} \sum_{j=0}^{J} \Delta u_i \int_{v_j}^{v_{j+1}} \left(\frac{\partial C_{ex}}{\partial u}\right)^2 dv$$

$$= \sum_{i=0}^{I} \sum_{j=0}^{J} \frac{1}{\Delta u_i} \int_{v_j}^{v_{j+1}} \left(M_{i,j}\frac{v-v_j}{\Delta v_j} - N_{i,j}\right)^2 dv$$

$$= \sum_{i=0}^{I} \sum_{j=0}^{J} \frac{1}{3}\frac{\Delta v_j}{\Delta u_i}(M_{i,j}^2 - 3M_{i,j}N_{i,j} + 3N_{i,j}^2)$$

$$= \sum_{i=0}^{I} \sum_{j=0}^{J} \frac{1}{3}\frac{\Delta v_j}{\Delta u_i}[(N_{i,j})^2 + N_{i,j}N_{i,j+1} + (N_{i,j+1})^2]$$

$$\leq \frac{2}{3}\sum_{j=0}^{J} \Delta v_j v_j + \frac{1}{3}\sum_{j=0}^{J} \Delta v_j v_{j+1}$$

$$= \sum_{j=0}^{J} \Delta v_j v_j + \frac{1}{3}\sum_{j=0}^{J} (\Delta v_j)^2$$

$$\leq \frac{1}{2}$$

where $C_{i,j} = C(u_i, v_j)$, $M_{i,j} = C_{i,j} - C_{i,j+1} - C_{i+1,j} + C_{i+1,j+1}$, $N_{i,j} = C_{i,j} - C_{i+1,j}$.

Similarly,

$$\int_0^1 \int_0^1 \left(\frac{\partial C_{ex}}{\partial v}\right)^2 dudv \leq \frac{2}{3}\sum_{i=0}^{I} \Delta u_i u_i + \frac{1}{3}\sum_{i=0}^{I} \Delta u_i u_{i+1} \leq \frac{1}{2}$$

Therefore, $|C_{ex}| \leq 1$ and it is strictly less than 1 for finite discrete random variables.

Example 4.3 Let X has the following distribution and $Y = X^2$.

Table 4-6 Distribution of X

$X = x$	1	2	3	4	5
$P(X = x)$	1/8	1/8	1/2	1/8	1/8

基于Copula的相关性测度
Measures of Association and Dependence Through Copulas

Table 4-7 Subcopula of X and Y

$V=v$ \ $U=u$	1/8	1/4	3/4	7/8	1
1/8	1/8	1/8	1/8	1/8	1/8
1/4	1/8	1/4	1/4	1/4	1/4
3/4	1/8	1/4	3/4	3/4	3/4
7/8	1/8	1/4	3/4	7/8	7/8
1	1/8	1/4	3/4	7/8	1

Then the subcopula associated with $H_{X,Y}$ is in Table 4-3.

After calculation, we obtain $\|C\|^2 = 1$ reaches its upper bound and $\mu(C) = 1$. But if we use the bilinear extension method to extend subcopula C to a continuous copula C_{ex}, we will find the sobolev norm $|C_{ex}|^2 = 0.896$, which does not reaches the upper bound 1. So C_{ex} is not a MCD.

4.4 Remarks on measures of dependence

Pearson's r measures linear relationship, Spearman's ρ and Kendall's τ measure concordance, our definition $\mu_{\frac{1}{2}}(Y|X)$ and $\mu_{\frac{1}{2}}(X|Y)$ can be used to measure a relationship which is nonlinear and discordant. For illustration, an example is given below. Before comparing those measures, an introduction to Spearman's ρ and Kendall's τ for discrete variables which are given by Nešlehová is needed.

Definition 4.7 Let X and Y are discrete random variables with subcopula C and the corresponding bilinear extension C_{ex}, then

(1) the non-continuous version of Kendall's tau is given by

$$\tau(X,Y) = \frac{4\int C_{ex} dC_{ex} - 1}{\sqrt{(1 - E(\Delta F_X(x)))(1 - E(\Delta G_Y(y)))}}$$

Measures for Discrete MCD and Functional Dependence

and

(2) the non-continuous version of Spearman's rho is given by

$$\rho(X,Y) = \frac{12 \int C_{ex} du dv - 3}{\sqrt{(1 - E(\Delta F_X(x))^2)(1 - E(\Delta G_Y(y))^2)}},$$

where $\Delta F_X(x) = P(X=x)$ and $\Delta G_Y(y) = P(Y=y)$.

Example 4.4 Table 4-8 is the joint p.d.f. of X and Y, Table 4-9 is the corresponding subcopula.

Table 4-8 Joint Distribution of X and Y

$Y=y$ \ $X=x$	-2	-1	0	1	2
0	1/64	1/128	7/32	0	1/128
1	1/64	15/64	0	1/8	0
4	7/64	0	1/64	0	1/4

We first calculate Pearson's coefficient r, Spearman's ρ and Kendall's τ:

$$r = 0.324, \ \rho = 0.249, \ \tau = 0.189$$

Table 4-9 Subcopula of X and Y

$V=v$ \ $U=u$	9/64	49/128	79/128	95/128	1
1/4	1/64	3/128	31/128	31/128	1/4
5/8	1/32	35/128	63/128	79/128	5/8
1	9/64	49/128	79/128	95/128	1

It is hardly to say X and Y have any kind of strong relation from the above values, while calculating $\mu_{\frac{1}{2}}(Y|X)$ and $\mu_{\frac{1}{2}}(X|Y)$ tells us a different thing. Here $\mu_{\frac{1}{2}}(Y|X) = 0.888$, $\mu_{\frac{1}{2}}(X|Y) = 0.538$. The large $\mu_{\frac{1}{2}}(Y|X)$ indicates Y depends strongly on X in the sense of functional relationship. Of

基于Copula的相关性测度
Measures of Association and Dependence Through Copulas

course, to find the exact functional relationship is a different story, but at least we know this strong relation exists.

4.5 Other measures

We will introduce several more possible measures. First, we define the support of a discrete random variable X to be the set supp $(X) = \{x \mid P(X=x) > 0\}$.

4.5.1 The measure μ_0^2

Definition 4.8 The normalized measre of dependence for discrete random variables X and Y with probability mass function f and g, respectively, is defined by

$$\mu_0^2(Y \mid X) = \frac{\sum_x \sum_y (P(Y \leqslant y \mid X = x) - G(y))^2 f(x) g(y)}{\sum_y G(y)(1 - G(y)) g(y)} \qquad (4-5)$$

where G is the marginal distribution function associated to Y.

The properties of the measure μ_0^2 in (4-5) are listed in the following theorem.

Theorem 4.8 Let X and Y be discrete random variables with probability mass function f and g repectively. The measure μ_0^2 has the following properties:

a. $0 \leqslant \mu_0^2(Y|X) \leqslant 1$;

b. $\mu_0^2(Y|X) = 0$ if and only if X and Y are independent;

c. $\mu_0^2(Y|X) = 1$ if and only if Y is completely dependent on X;

d. For any injective function ϕ: supp$(X) \mapsto \mathbb{R}$. $\mu_0^2(Y|\phi(X)) = \mu_0^2(Y|X)$;

e. For any strictly increasing function ψ: supp$(Y) \mapsto \mathbb{R}$, $\mu_0^2(\psi(Y)|X) = \mu_0^2(Y|X)$;

f. If the probatility mass function associated to Y is uniform, then $\mu_0^2(\psi(Y)|X) = \mu_0^2(Y|X)$ for any strictly decreasing function ψ: supp$(Y) \mapsto \mathbb{R}$.

The drawback of the properties of μ_0^2 is that it requires the condition in property f: the probability mass function associated to discrete random variable Y must

Measures for Discrete MCD and Functional Dependence

be uniform. If not, μ_0^2 is not invaraint under strictly decreasing transformation and then μ_0^2 is also not invaraint under strictly monotone transformation. The question is "How can we define the measure of dependence with the property f in Theorem 4.8 without its assumption". This question will be answered in the next section, we will define the new measure of complete dependence, called $\bar{\lambda}$, and also show the properties of $\bar{\lambda}$.

4.5.2 The measure $\bar{\lambda}$

Since μ_0^2 in (4-5) is not invariant under strictly decreasing transformation, we will define the new measure, called $\bar{\lambda}$, which will overcome the drawback of μ_0^2.

Definition 4.9 The normalized measure of dependence for finite random variables X and Y with probability mass function f and g, respectively, is defined by

$$\bar{\lambda}(Y \mid X) = \frac{\sum_x \sum_y (P(Y \leqslant y \mid X = x) - G(y))^2 f(x)}{\sum_y G(y)(1 - G(y))}$$

where G is the marginal distribution function associated to Y.

Note that $\bar{\lambda}$ is only defined for finite random variables, not discrete, since $\bar{\lambda}$ only converges if we sum up over a finite number of y. This implies that the support of Y must be a finite set. Observe that the formula of $\bar{\lambda}$ in this definition is similar to the definition of μ_0^2. The difference is just the formula of $\bar{\lambda}$ is defined without the probability mass function $g(y)$. The other aspects of $\bar{\lambda}$ can be obtained analogously to the measure μ_0^2 except the invariant under strictly decreasing transformation. The consequences can be summarized as follows.

Theorem 4.9 Let X and Y be finite random variables with probability mass function f and g, respectively. The measure $\bar{\lambda}$ has the following properties:

a. $0 \leqslant \lambda(Y|X) \leqslant 1$;

b. $\bar{\lambda}(Y|X) = 0$ if and only if X and Y are independent;

基于Copula的相关性测度
Measures of Association and Dependence Through Copulas

    c. $\overline{\lambda}(Y|X) = 1$ if and only if Y is completely dependent on X;

    d. For any injective function ϕ: $\mathrm{supp}(X) \mapsto \mathbb{R}$, $\overline{\lambda}(Y|\phi(X)) = \overline{\lambda}(Y|X)$;

    e. For any strictly monotone function ψ: $\mathrm{supp}(Y) \mapsto \mathbb{R}$, $\overline{\lambda}(\psi(Y)|X) = \overline{\lambda}(Y|X)$.

Although the measure $\overline{\lambda}$ has the advantage properties over μ_0^2, there is one defect of the measure $\overline{\lambda}$. Comparing the definition of $\overline{\lambda}$ with μ_0^2, we see that the measure $\overline{\lambda}$ is only defined for finite random variables but the measure μ_0^2 is defined for discrete random variables.

The next theorem will answer the question "When does the value of μ_0^2 is equal to $\overline{\lambda}$".

Theorem 4.10 Let X and Y be finite random variables with probability mas function f and g, respectively. If $|\mathrm{supp}(Y)| = 2$, then $\mu_0^2(Y|X) = \overline{\lambda}(Y|X)$.

Now we can define a new measure of MCD for finite random variables.

Definition 4.10 The measure of mutual complete dependence for discrete random variables X and Y with probability mass functions f and g, respectively, is defined by

$$\overline{\lambda}(X, Y) = \frac{1}{2}(\overline{\lambda}(X|Y) + \overline{\lambda}(Y|X))$$

The properties of the measure $\overline{\lambda}(X, Y)$ are given by the following theorem.

Theorem 4.11 The measure $\overline{\lambda}(X, Y)$ has the following properties.

    a. $\overline{\lambda}(X, Y) = \overline{\lambda}(Y, X)$;

    b. $0 \leqslant \overline{\lambda}(X, Y) \leqslant 1$;

    c. $\overline{\lambda}(X, Y) = 0$ if and only if X and Y are independent;

    d. $\overline{\lambda}(X, Y) = 1$ if and only if X and Y are mutually completely dependent.

Measures for Discrete MCD and Functional Dependence

4.5.3 Construction of the measure

This section shows the way to construct and define the measure $\bar{\lambda}$. Moreover, we will discuss several aspects of the measure $\bar{\lambda}$.

Definition 4.11 For any discrete random variables X and Y, define

$$\lambda(Y \mid X) = \sum_x \sum_y \left(P(Y \leq y \mid X = x) - \frac{1}{2} \right)^2 P(X = x)$$

where x, y ranges over the support of X and Y, respectively.

Note that the support of Y must be a finite set since λ only converges if we sum up over a finite number of y, so $\lambda(Y \mid X)$ is only well defined when supp (Y) is finite. Next, we will discuss some properties of $\bar{\lambda}$. One interesting question is "what is the range of $\lambda(Y \mid X)$". First, let us give some notations.

Definition 4.12 Let f and g be probability mass functions. The maximum of the function λ is defined by

$$\lambda_{f,g}^{\max} = \max \left\{ \lambda(Y \mid X) \;\middle|\; \begin{array}{l} X \text{ and } Y \text{ are finte random variables with} \\ \text{probability mass function } f \text{ and } g, \text{ respectively} \end{array} \right\}$$

and the minimum of the function λ is defined by

$$\lambda_{f,g}^{\min} = \min \left\{ \lambda(Y \mid X) \;\middle|\; \begin{array}{l} X \text{ and } Y \text{ are finte random variables with} \\ \text{probability mass function } f \text{ and } g, \text{ respectively} \end{array} \right\}$$

Theorem 4.12 Let X and Y be finite random variables with probability mass function f and g, respectively. The following statements hold:

a. $\lambda_{f,g}^{\max} \leq \dfrac{|\text{supp}(Y)|}{4}$;

b. $\lambda(Y \mid X) = \dfrac{|\text{supp}(Y)|}{4}$ if and only if there is a function φ such that $Y = \varphi(X)$ a.s.;

c. $\lambda(Y \mid X) = \lambda_{f,g}^{\min}$ if and only if X and Y are independent.

基于Copula的相关性测度
Measures of Association and Dependence Through Copulas

The statement b reveals that the maximum value of λ is exactly $\dfrac{|\operatorname{supp}(Y)|}{4}$. This result follows from the definition of λ. Let X and Y be finite random variables with probability mass function f and g, respectively. Assume that $P(Y \leqslant y | X=x) \in \{0, 1\}$ for all $x \in \operatorname{supp}(X)$ and $y \in \operatorname{supp}(Y)$. Then $|P(Y \leqslant y|X=x) - \dfrac{1}{2}| = \dfrac{1}{2}$ and hence $(P(Y \leqslant y|X=x) - \dfrac{1}{2})^2 = \dfrac{1}{4}$. Since we define the function λ without probability mass function g, then we must sum up the constant $\dfrac{1}{4}$ over all numbers of $y \in \operatorname{supp}(Y)$, i.e., we sum up the constant $\dfrac{1}{4}$ for $|\operatorname{supp}(Y)|$ times. This implies the maximum value of λ is $\dfrac{|\operatorname{supp}(Y)|}{4}$.

Theorem 4.13 Let X and Y be finite random variables with probability mass function f and g, respectively. The function λ has the following properties:

a. For any injective function ϕ: $\operatorname{supp}(X) \to \mathbb{R}$, $\lambda(Y|\phi(X)) = \lambda(Y|X)$;

b. For any strictly monotone function ψ: $\operatorname{supp}(Y) \to \mathbb{R}$, $\lambda(\psi(Y)|X) = \lambda(Y|X)$.

From this theorem, we can see that λ has the good property for decreasing transformation over ω. Now, all aspects of λ that we have proved are sufficient to construct the new measure of dependence. Let us begin with the normalization of the function λ into the measure $\bar{\lambda}$.

Definition 4.13 The normalized measure of dependence for finite random variables X and Y with probability mass function f and g, respectively, is defined by

$$\bar{\lambda}(Y \mid X) = \frac{\lambda(Y \mid X) - \lambda(Y_0 \mid X_0)}{\frac{|\operatorname{supp}(Y)|}{4} - \lambda(Y_0 \mid X_0)}$$

$$= \frac{\sum_x \sum_y \left(P(Y \leqslant y \mid X = x) - G(y)\right)^2 f(x)}{\sum_y G(y)(1 - G(y))}$$

Measures for Discrete MCD and Functional Dependence

where X_0 and Y_0 are independent finite random variables with probability mass functions f and g, respectively.

This result follows from Lemma 4.6, Lemma 4.7, and Lemma 4.8 in Section 4.5.4. Moreover, the properties of $\overline{\lambda}$ can be obtained by the properties of λ via normalization. The consequences can be summarized into the following theorem.

Theorem 4.14 Let X and Y be finite random variables with probability mass function f and g, respectively. The measure $\overline{\lambda}$ has the following properties:

a. $0 \leq \overline{\lambda}(Y|X) \leq 1$;

b. $\overline{\lambda}(Y|X) = 0$ if and only if X and Y are independent;

c. $\overline{\lambda}(Y|X) = 1$ if and only if there is a function φ such that $Y = \varphi(X)$ a.s.;

d. For any injective function ϕ: $\mathrm{supp}(X) \to \mathbb{R}$, $\overline{\lambda}(Y|\phi(X)) = \overline{\lambda}(Y|X)$;

e. For any strictly monotone function ψ: $\mathrm{supp}(Y) \to \mathbb{R}$, $\overline{\lambda}(\psi(Y)|X) = \overline{\lambda}(Y|X)$.

As a result, we can see that although the measure $\overline{\lambda}$ can improve the properties of the measure μ_0^2, there is one defect of $\overline{\lambda}$; the measure $\overline{\lambda}$ is only defined for finite random variable X and Y. Hence the measure $\overline{\lambda}$ is not appropriate for discrete random variables.

4.5.4 Proofs of the construction of $\overline{\lambda}$

Theorem 4.15 Let X and Y be finite random variables with probability mass functions f and g, respectively. Then $\lambda_{f,g}^{\max} \leq \dfrac{|\mathrm{supp}(Y)|}{4}$.

Proof: Let $\widetilde{X}$ and $\widetilde{Y}$ be finite random variables with probability mass functions f and g, respectively. Assume that either $P(\widetilde{Y} \leq \widetilde{y} | \widetilde{X} = \widetilde{x}) = 0$ or $P(\widetilde{Y} \leq \widetilde{y} | \widetilde{X} = \widetilde{x}) = 1$ for all $\widetilde{x} \in \mathrm{supp}(\widetilde{X})$ and $\widetilde{y} \in \mathrm{supp}(\widetilde{Y})$. Since $0 \leq P(Y \leq y | X = x) \leq 1$,

$$\lambda(Y|X) = \sum_x \sum_y \left(P(Y \leq y | X = x) - \frac{1}{2}\right)^2 P(X = x)$$

$$\leq \sum_{\widetilde{x}} \sum_{\widetilde{y}} \left(P(\widetilde{Y} \leq \widetilde{y} | \widetilde{X} = \widetilde{x}) - \frac{1}{2}\right)^2 P(\widetilde{X} = \widetilde{x})$$

$$= \frac{|\mathrm{supp}(Y)|}{4}$$

基于Copula的相关性测度
Measures of Association and Dependence Through Copulas

which implies $\lambda_{f,g}^{\max} \leq \dfrac{|\operatorname{supp}(Y)|}{4}$.

Theorem 4.16 For any finite random variables X and Y, $\lambda(Y|X) = \dfrac{|\operatorname{supp}(Y)|}{4}$ if and only if there is a function φ such that $Y = \varphi(X)$ a.s.

Proof: Assume that $\lambda(Y|X) = \dfrac{|\operatorname{supp}(Y)|}{4}$. Since $0 \leq P(Y \leq y|X=x) \leq 1$, $|P(Y \leq y|X=x) - \dfrac{1}{2}| \leq \dfrac{1}{2}$. The only way that $\lambda(Y|X) = \dfrac{|\operatorname{supp}(Y)|}{4}$ is $(P(Y \leq y|X=x) - \dfrac{1}{2})^2 = \dfrac{1}{4}$ for all $x \in \operatorname{supp}(X)$ and $y \in \operatorname{supp}(Y)$. Thus we have $|P(Y \leq y|X=x) - \dfrac{1}{2}| = \dfrac{1}{2}$, i.e., either $P(Y \leq y|X=x) = 0$ or $P(Y \leq y|X=x) = 1$. The goal is to find a function φ such that $P(Y=\varphi(X)) = 1$. Let $\varphi(x) = \min\{y \in \operatorname{supp}(Y) | P(Y \leq y|X=x) = 1\}$. Then $P(Y \leq y|X=x) = 1$ if and only if $y \geq \varphi(x)$. For any $w > \varphi(x)$, we have

$$P(Y=w|X=x) \leq P(\varphi(x) < Y \leq w|X=x)$$
$$= P(Y \leq w|X=x) - P(Y \leq \varphi(x)|X=x)$$
$$= 1 - 1 = 0$$

Since $P(Y \leq \varphi(x)|X=x) = 1$ and either $P(Y \leq y|X=x) = 0$ or $P(Y \leq y|X=x) = 1$, $P(Y < \varphi(x)|X=x) = 0$ and hence

$$P(Y=\varphi(x)|X=x) = P(Y \leq \varphi(x)|X=x) - P(Y < \varphi(x)|X=x)$$
$$= 1 - 0 = 1$$

Therefore,
$$P(Y = \varphi(X)) = \sum_x P(Y = \varphi(x), X = x)$$
$$= \sum_x P(Y = \varphi(x) \mid X = x)P(X = x)$$
$$= \sum_x P(X = x) = 1$$

Conversely, we assume that there is a function φ such that $Y = \varphi(X)$ a.s. Then

Measures for Discrete MCD and Functional Dependence

$$\lambda(Y \mid X) = \lambda(\varphi(X) \mid X)$$

$$= \sum_x \sum_y \left(P(\varphi(X) \leqslant y \mid X = x) - \frac{1}{2} \right)^2 P(X = x)$$

$$= \sum_x \sum_y \left(P(\varphi(x) \leqslant y) - \frac{1}{2} \right)^2 P(X = x)$$

It is easy to see that $P(\varphi(x) \leqslant y)$ is equal to 1 if $\varphi(x) \leqslant y$ and is equal to 0 if $\varphi(x) > y$. Hence, either $P(\varphi(x) \leqslant y) = 0$ or $P(\varphi(x) \leqslant y) = 1$, i.e., $\mid P(\varphi(x) \leqslant y) - \frac{1}{2} \mid = \frac{1}{2}$. Therefore, $\sum_y (P(\varphi(x) \leqslant y) - \frac{1}{2})^2 = \frac{\mid \text{supp}(Y) \mid}{4}$ which implies $\lambda(Y \mid X) = \frac{\mid \text{supp}(Y) \mid}{4}$.

Theorem 4.17 For any finite random variables X and Y associate to probability mass functions f and g, $\lambda(Y \mid X) = \lambda_{f,g}^{\min}$ if and only if X and Y are independent.

Proof: Let X_0 and Y_0 be independent finite random variables with probability mass function f and g, respectively. Then $P(Y_0 \leqslant y \mid X_0 = x) = P(Y_0 \leqslant y)$ and hence

$$\lambda(Y_0 \mid X_0) = \sum_x \sum_y \left(P(Y_0 \leqslant y) - \frac{1}{2} \right)^2 P(X_0 = x)$$

$$= \sum_x \sum_y \left(G(y) - \frac{1}{2} \right)^2 f(x)$$

Assume that finite random variables X and Y are not independent. Let $\Phi: [0, 1] \mapsto \mathbb{R}$ be a function defined by

$$\Phi(\alpha) = \sum_x \sum_y \left(\alpha P(Y \leqslant y \mid X = x) + (1-\alpha) P(Y_0 \leqslant y) - \frac{1}{2} \right)^2 P(X = x)$$

$$= \sum_x \sum_y \left(\alpha P(Y \leqslant y \mid X = x) + (1-\alpha) G(y) - \frac{1}{2} \right)^2 f(x)$$

Observe that $\Phi(0) = \lambda(Y_0 \mid X_0)$ and $\Phi(1) = \lambda(Y \mid X)$. Consider the first one-side derivative

$$\Phi'(\alpha) = 2 \sum_x \sum_y \left(\alpha P(Y \leqslant y \mid X = x) + (1-\alpha) G(y) - \frac{1}{2} \right)$$

$$\left(P(Y \leqslant y \mid X = x) - G(y) \right) f(x)$$

基于Copula的相关性测度
Measures of Association and Dependence Through Copulas

Then

$$\Phi'(0) = 2\sum_x \sum_y \left(G(y) - \frac{1}{2}\right)\left(P(Y \leqslant y \mid X = x) - G(y)\right)f(x)$$

$$= 2\sum_x \sum_y [G(y)P(Y \leqslant y \mid X = x) - (G(y))^2] f(x) -$$

$$2\sum_x \sum_y \left[\frac{1}{2}P(Y \leqslant y \mid X = x) - \frac{1}{2}G(y)\right] f(x)$$

$$= 2\sum_x \sum_y [G(y)P(Y \leqslant y, X = x) - (G(y))^2 f(x)] -$$

$$2\sum_x \sum_y \left[\frac{1}{2}P(Y \leqslant y, X = x) - \frac{1}{2}G(y)f(x)\right]$$

$$= 2\left(\sum_y (G(y))^2 - \sum_y (G(y))^2 - \frac{1}{2}\sum_y G(y) + \frac{1}{2}\sum_y G(y)\right)$$

$$= 0$$

Thus, $\alpha = 0$ is a critical point of a function Φ. Next, we will show that a point $\alpha = 0$ minimizes Φ. By the second derivative test,

$$\Phi''(\alpha) = 2\sum_x \sum_y \left(P(Y \leqslant y \mid X = x) - G(y)\right)^2 f(x)$$

and hence

$$\Phi''(0) = 2\sum_x \sum_y \left(P(Y \leqslant y \mid X = x) - G(y)\right)^2 f(x)$$

Since X and Y are not independent, there are $x \in$ supp (X) and $y \in$ supp (Y) such that $|P(Y \leqslant y | X = x) - G(y)| > 0$, i.e., $\Phi''(0) > 0$. Thus, $\Phi(0)$ is a minimum value of a function Φ, i.e., $\Phi(0) \leqslant \Phi(1)$. Therefore, $\lambda(Y_0 | X_0) \leqslant \lambda(Y|X)$ which implies $\lambda(Y_0|X_0) = \lambda_{f,g}^{\min}$.

Proof of Theorem 4.12: Follow from Theorem 4.15, Theorem 4.16, and Theorem 4.17, the proof is completed.

Proof of Theorem 4.13: Let X and Y be finite random variables with probability mass function f and g, respectively. Now, we distinguish the proof into the three parts.

First, let ϕ: supp $(X) \mapsto \mathbb{R}$ be an injective function. Then there is an in-

Measures for Discrete MCD and Functional Dependence

verse function ϕ^{-1}: Ran $\phi \mapsto \text{supp}(X)$ such that $\phi^{-1}(\phi(x)) = x$ for any $x \in \text{supp}(X)$.

Denote $W = \phi(X)$. Therefore,

$$\lambda(Y \mid W) = \sum_w \sum_y \left(P(Y \leqslant y \mid W = w) - \frac{1}{2}\right)^2 P(W = w)$$

$$= \sum_{w \in \text{supp}(W)} \sum_y \left(P(Y \leqslant y \mid \phi(X) = w) - \frac{1}{2}\right)^2 P(\phi(X) = w)$$

$$= \sum_{w \in \text{supp}(W)} \sum_y \left(P(Y \leqslant y \mid X = \phi^{-1}(w)) - \frac{1}{2}\right)^2 P(X = \phi^{-1}(w))$$

$$= \sum_{x \in \text{supp}(X)} \sum_y \left(P(Y \leqslant y \mid X = x) - \frac{1}{2}\right)^2 P(X = x)$$

$$= \lambda(Y \mid X)$$

which implies λ is invariant under injective transformation of finite random variable X.

Second, let $\psi : \text{supp}(Y) \mapsto \mathbb{R}$ be a strictly increasing function. Then there is an inverse function $\psi^{-1} : \text{Ran}\,\psi \mapsto \text{supp}(Y)$ such that $\psi^{-1}(\psi(y)) = y$ for any $y \in \text{supp}(Y)$. Since ψ is strictly increasing, $\psi(u) \leqslant v$ if and only if $\psi^{-1}(\psi(u)) \leqslant \psi^{-1}(v)$, i.e., $\psi(u) \leqslant v$ if and only if $u \leqslant \psi^{-1}(v)$. Denote $Z = \psi(Y)$. Then

$$\lambda(Z \mid X) = \sum_x \sum_z \left(P(Z \leqslant z \mid X = x) - \frac{1}{2}\right)^2 P(X = x)$$

$$= \sum_x \sum_{z \in \text{supp}(Z)} \left(P(\psi(Y) \leqslant z \mid X = x) - \frac{1}{2}\right)^2 f(x)$$

$$= \sum_x \sum_{z \in \text{supp}(Z)} \left(P(Y \leqslant \psi^{-1}(z) \mid X = x) - \frac{1}{2}\right)^2 f(x)$$

$$= \sum_x \sum_{y \in \text{supp}(Y)} \left(P(Y \leqslant y \mid X = x) - \frac{1}{2}\right)^2 f(x)$$

$$= \lambda(Y \mid X)$$

基于Copula的相关性测度
Measures of Association and Dependence Through Copulas

Lastl, let $\psi:$ supp $(Y) \mapsto R$ be a strictly decreasing function. Since supp (Y) is a finite set, we assume that $\mathrm{supp}(Y) = \{y_1, y_2, \cdots, y_n\}$ where $y_1 < y_2 < \cdots < y_n$. Denote $Z = \psi(Y)$ and mention that for strictly decreasing function ψ, we have $\psi(u) \leqslant v$ if and only if $u \geqslant \psi^{-1}(v)$. Hence,

$$\lambda(Z \mid X) = \sum_x \sum_z \left(P(Z \leqslant z \mid X = x) - \frac{1}{2} \right)^2 P(X = x)$$

$$= \sum_x \sum_{z \in \mathrm{supp}(Z)} \left(P(\psi(Y) \leqslant z \mid X = x) - \frac{1}{2} \right)^2 f(x)$$

$$= \sum_x \sum_{z \in \mathrm{supp}(Z)} \left(P(Y \geqslant \psi^{-1}(z) \mid X = x) - \frac{1}{2} \right)^2 f(x)$$

$$= \sum_x \sum_{y \in \mathrm{supp}(Y)} \left(P(Y \geqslant y \mid X = x) - \frac{1}{2} \right)^2 f(x)$$

$$= \sum_x \sum_{i=1}^n \left(P(Y \geqslant y_i \mid X = x) - \frac{1}{2} \right)^2 f(x)$$

$$= \sum_x \sum_{i=1}^n \left(1 - P(Y < y_i \mid X = x) - \frac{1}{2} \right)^2 f(x)$$

$$= \sum_x \sum_{i=1}^n \left(\frac{1}{2} - P(Y \leqslant y_{i-1} \mid X = x) \right)^2 f(x)$$

$$= \sum_x \left(\frac{1}{4} + \sum_{i=2}^n \left(\frac{1}{2} - P(Y \leqslant y_{i-1} \mid X = x) \right)^2 \right) f(x)$$

$$= \frac{1}{4} + \sum_x \sum_{j=1}^{n-1} \left(\frac{1}{2} - P(Y \leqslant y_j \mid X = x) \right)^2 f(x)$$

From the fact that $P(Y \leqslant y_n \mid X = x) = 1$ for all $x \in$ supp (X),

$$\left(\frac{1}{2} P(Y \leqslant y_n \mid X = x) \right)^2 = \frac{1}{4}$$

for all $x \in$ supp (X) and hence

$$\sum_x \left(\frac{1}{2} - P(Y \leqslant y_n \mid X = x) \right)^2 f(x) = \frac{1}{4}$$

Thus,

Measures for Discrete MCD and Functional Dependence

$\lambda(Z \mid X)$

$$= \sum_x \left(\frac{1}{2} - P(Y \leqslant y_n \mid X = x)\right)^2 f(x) + \sum_x \sum_{j=1}^{n-1} \left(\frac{1}{2} - P(Y \leqslant y_j \mid X = x)\right)^2 f(x)$$

$$= \sum_x \left(\left(\frac{1}{2} - P(Y \leqslant y_n \mid X = x)\right)^2 + \sum_{j=1}^{n-1} \left(\frac{1}{2} - P(Y \leqslant y_j \mid X = x)\right)^2 \right) f(x)$$

$$= \sum_x \sum_{j=1}^n \left(\frac{1}{2} - P(Y \leqslant y_j \mid X = x)\right)^2 f(x)$$

$=\lambda(Y \mid X)$

The two previous parts show that λ is invariant under strictly increasing and strictly decreasing transformation which implies λ is invariant under strictly monotone transformation of finite random variable Y. Therefore, the proof is completed.

A few following lemmas are important for normalization.

Lemma 4.6 For any finite random variables X and Y and independent finite random variables X_0 and Y_0,

$$\lambda(Y \mid X) - \lambda(Y_0 \mid X_0) = \sum_x \sum_y \left(P(Y \leqslant y \mid X = x)^2 - G(y)^2\right) f(x)$$

Proof: Using the fact that $\sum_x \sum_y P(Y \leqslant y \mid X = x) f(x) = \sum_y G(y)$,

$\lambda(Y \mid X) - \lambda(Y_0 \mid X_0)$

$$= \sum_x \sum_y \left(\left(P(Y \leqslant y \mid X = x) - \frac{1}{2}\right)^2 - \left(G(y) - \frac{1}{2}\right)^2\right) f(x)$$

$$= \sum_x \sum_y \left(P(Y \leqslant y \mid X = x)^2 - P(Y \leqslant y \mid X = x) + G(y) - G(y)^2\right) f(x)$$

$$= \sum_x \sum_y \left(P(Y \leqslant y \mid X = x)^2 - G(y)^2\right) f(x) -$$

$$\sum_x \sum_y P(Y \leqslant y \mid X = x) f(x) + \sum_x \sum_y G(y) f(x)$$

$$= \sum_x \sum_y \left(P(Y \leqslant y \mid X = x)^2 - G(y)^2\right) f(x) - \sum_y G(y) + \sum_y G(y)$$

$$= \sum_x \sum_y \left(P(Y \leqslant y \mid X = x)^2 - G(y)^2\right) f(x)$$

Lemma 4.7 For any finite random variables X and Y,

基于Copula的相关性测度
Measures of Association and Dependence Through Copulas

$$\sum_x \sum_y \left(P(Y \leqslant y \mid X = x) - G(y)\right)^2 f(x)$$

$$= \sum_x \sum_y \left(P(Y \leqslant y \mid X = x)^2 - G(y)^2\right) f(x)$$

Proof: From the fact that $\sum_x \sum_y P(Y \leqslant y \mid X = x) G(y) f(x) = \sum_y G(y)^2$, we have

$$\sum_x \sum_y \left(P(Y \leqslant y \mid X = x) - G(y)\right)^2 f(x)$$

$$= \sum_x \sum_y \left(P(Y \leqslant y \mid X = x)^2 - 2P(Y \leqslant y \mid X = x) G(y) + G(y)^2\right) f(x)$$

$$= \sum_x \sum_y \left(P(Y \leqslant y \mid X = x)^2\right) f(x) -$$

$$2 \sum_x \sum_y P(Y \leqslant y \mid X = x) G(y) f(x) + \sum_x \sum_y G(y)^2 f(x)$$

$$= \sum_x \sum_y \left(P(Y \leqslant y \mid X = x)^2\right) f(x) - 2 \sum_y G(y)^2 + \sum_y G(y)^2$$

$$= \sum_x \sum_y \left(P(Y \leqslant y \mid X = x)^2\right) f(x) - \sum_y G(y)^2$$

$$= \sum_x \sum_y \left(P(Y \leqslant y \mid X = x)^2 - G(y)^2\right) f(x)$$

Lemma 4.8 For any independent finite random variables X_0 and Y_0,

$$\frac{|\mathrm{supp}(Y)|}{4} - \lambda(Y_0 \mid X_0) = \sum_y G(y)(1 - G(y))$$

Proof: This result follows from simple computation

$$\frac{|\mathrm{supp}(Y)|}{4} - \lambda(Y_0 \mid X_0) = \frac{|\mathrm{supp}(Y)|}{4} - \sum_x \sum_y \left(G(y) - \frac{1}{2}\right)^2 f(x)$$

$$= \frac{|\mathrm{supp}(Y)|}{4} - \sum_y \left(G(y)^2 - G(y) + \frac{1}{4}\right)$$

$$= \frac{|\mathrm{supp}(Y)|}{4} + \sum_y \left(G(y) - G(y)^2\right) - \frac{1}{4}\sum_y$$

$$= \frac{|\mathrm{supp}(Y)|}{4} + \sum_y G(y)(1 - G(y)) - \frac{|\mathrm{supp}(Y)|}{4}$$

$$= \sum_y G(y)(1 - G(y))$$

Measures for Discrete MCD and Functional Dependence

Proof of Theorem 4.14: Every property follows from all aspects of λ via normalization.

Proof of Theorem 4.11: See Proof of Theorem 4.14.

Proof of Theorem 4.10: Assume that supp $(Y) = \{y_1, y_2\}$. Then

$\bar{\omega}(Y \mid X)$

$= \dfrac{\sum_x \sum_y (P(Y \leqslant y \mid X = x) - G(y))^2 f(x)g(y)}{\sum_y G(y)(1 - G(y))g(y)}$

$= \dfrac{\sum_x \sum_{i=1}^{2} (P(Y \leqslant y_i \mid X = x) - G(y_i))^2 f(x)g(y_i)}{\sum_{i=1}^{2} G(y_i)(1 - G(y_i))g(y_i)}$

$= \dfrac{\sum_x (P(Y \leqslant y_1 \mid X = x) - G(y_1))^2 f(x)g(y_1)}{G(y_1)(1 - G(y_1))g(y_1) + G(y_2)(1 - G(y_2))g(y_2)}$

$+ \dfrac{\sum_x (P(Y \leqslant y_2 \mid X = x) - G(y_2))^2 f(x)g(y_2)}{G(y_1)(1 - G(y_1))g(y_1) + G(y_2)(1 - G(y_2))g(y_2)}$

From the fact that $P(Y \leqslant y_2 \mid X = x) = G(y_2) = 1$, we have

$\bar{\omega}(Y \mid X)$

$= \dfrac{\sum_x (P(Y \leqslant y_1 \mid X = x) - G(y_1))^2 f(x)g(y_1)}{G(y_1)(1 - G(y_1))g(y_1)}$

$= \dfrac{\sum_x (P(Y \leqslant y_1 \mid X = x) - G(y_1))^2 f(x)}{G(y_1)(1 - G(y_1))}$

$= \dfrac{\sum_x ((P(Y \leqslant y_1 \mid X = x) - G(y_1))^2 f(x) + (P(Y \leqslant y_2 \mid X = x) - G(y_2))^2 f(x))}{G(y_1)(1 - G(y_1)) + G(y_2)(1 - G(y_2))}$

$= \dfrac{\sum_x \sum_{i=1}^{2} (P(Y \leqslant y_i \mid X = x) - G(y_i))^2 f(x)}{\sum_{i=1}^{2} G(y_i)(1 - G(y_i))}$

$= \dfrac{\sum_x \sum_y (P(Y \leqslant y \mid X = x) - G(y))^2 f(x)}{\sum_y G(y)(1 - G(y))}$

$= \bar{\lambda}(Y \mid X)$

Therefore, the proof is completed.

5
Nonparametric Estimation of the Measure of Functional Dependence

From Definition 2.12, we know that estimating the measure of MCD is essentially estimating the norm of a copula. The Sobolev norm of a copula C given by reference is:

$$\|C\|^2 = \int_0^1 \int_0^1 \left[\left(\frac{\partial C}{\partial u}\right)^2 + \left(\frac{\partial C}{\partial v}\right)^2\right] du dv$$

which can also be written as:

$$\begin{aligned}\|C\|^2 &= \int_0^1 \int_0^1 \left[\left(\frac{\partial C}{\partial u}\right)^2 + \left(\frac{\partial C}{\partial v}\right)^2\right] du dv \\ &= \int_0^1 \int_0^1 \left[\left(\int_0^v c(u,y) dy\right)^2 + \left(\int_0^u c(x,v) dx\right)^2\right] du dv\end{aligned} \quad (5-1)$$

where c is the density of the copula. This formula provides two possible ways to estimate the norm, one is through copula, the other one is through density of copula, both of them will be computational expensive. For simplicity, we restrict to the case $d=2$, and start from the most convenient situation, in which we assume that the copula has continuous partial derivatives. Consider an independent and identically distributed sample $(X_1, Y_1), \cdots, (X_n, Y_n)$ of a bivariate random vector (X, Y) with joint distribution function H and marginal distribution functions F and G.

Nonparametric Estimation of the Measure of Functional Dependence

5.1 Nonparametric estimation through the density of copula

Pseudo-observations are often used to estimate copulas, so we will first discuss why should we use pseudo-observations.

5.1.1 Estimating with pseudo-observations

Copula can be estimated via pseudo-observations $(\hat{F}(X_i), \hat{G}(Y_i))$, where $\hat{F}$ and $\hat{G}$ are the empirical distribution functions

$$\hat{F}(x) = \frac{1}{n}\sum_{i=1}^{n}\mathbf{1}(X_i \leqslant x) \quad \text{and} \quad \hat{G}(y) = \frac{1}{n}\sum_{i=1}^{n}\mathbf{1}(Y_i \leqslant y)$$

In some situations, using pseudo-observations will reduce the variance of estimators. To illustrate this, let's consider a point $(u, v) \in I^2$ with neighborhood $[u-h, u+h] \times [v-h, v+h]$, which lies in the interior of the unit square.

Assume that the margins are known, or equivalently, let $(U_1, V_1), \cdots, (U_n, V_n)$ be a sample from a copula. The number of points in the small square, say N_1 is following binomial distribution, denoted by $N_1 \sim B(n, p_1)$ with

$$p_1 = P(U, V) \in [u-h, u+h] \times [v-h, v+h]$$
$$= C(u+h, u+h) + C(u-h, v-h) - C(u-h, v+h) - C(u+h, v-h)$$
$$Var(N_1) = np_1(1-p_1)$$

If the margins are unknown, or equivalently, $(\hat{U}_1, \hat{V}_1), \cdots, (\hat{U}_n, \hat{V}_n)$ are pseudoobservations from a copula, then consider the same neighborhood of (u, v).

$$\hat{U}_i \in [u-h, u+h] = \lfloor 2hn \rfloor \quad \text{or} \quad \lfloor 2hn+1 \rfloor$$

where $\lfloor \ \rfloor$ denotes the integer part. The number of points in the same square is denoted by N_2. $N_2 \sim B(\lfloor 2hn \rfloor, p_2)$ or $N_2 \sim B(\lfloor 2hn+1 \rfloor, p'_2)$, where

基于Copula的相关性测度
Measures of Association and Dependence Through Copulas

$$p_2 = P((\hat{U}, \hat{V}) \in [u-h, u+h] \times [v-h, v+h] \mid \hat{U} \in [u-h, u+h])$$
$$= \frac{P((\hat{U}, \hat{V}) \in [u-h, u+h] \times [v-h, v+h])}{P(\hat{U} \in [u-h, u+h])}$$
$$\approx \frac{C(u+h, u+h) + C(u-h, v-h) - C(u-h, v+h) - C(u+h, v-h)}{2h}$$
$$= \frac{p_1}{2h}$$

and

$$p'_2 = P((\hat{U}, \hat{V}) \in [u-h, u+h] \times [v-h, v+h] \mid \hat{U} \in [u-h, u+h])$$
$$= \frac{P((\hat{U}, \hat{V}) \in [u-h, u+h] \times [v-h, v+h])}{P(\hat{U} \in [u-h, u+h])}$$
$$\approx \frac{C(u+h, u+h) + C(u-h, v-h) - C(u-h, v+h) - C(u+h, v-h)}{2h + 1/n}$$
$$= \frac{p_1}{2h + 1/n}$$

with both approximation hold only if the copula function of H is evenly distributed.

Then the expected number of observations is the same for both methods.
$$Var(N_2) \approx 2hnp_2(1-p_2) = 2hn\frac{p_1}{2h}(1 - \frac{p_1}{2h}) = \frac{n}{2h}p_1(2h - p_1)$$

or
$$Var(N_2) \approx (2hn+1)p'_2(1-p'_2)$$
$$= (2hn+1)\frac{p_1}{2h+1/n}(1 - \frac{p_1}{2h+1/n})$$
$$= \frac{np_1}{2h+1/n}(2h + 1/n - p_1)$$

Thus
$$\frac{Var(N_2)}{Var(N_1)} = \frac{np_1(2h - p_1)}{2hnp_1(1-p_1)} = \frac{2h - p_1}{2h - 2hp_1} \leq 1$$

or
$$\frac{Var(N_2)}{Var(N_1)} = \frac{np_1(2h + 1/n - p_1)}{(2h+1/n)np_1(1-p_1)} = \frac{(2h + 1/n - p_1)}{(2h + 1/n - (2h+1/n)p_1)} \leq 1$$

So if the copula function is evenly distributed, then the variance of the number of observations in the small square around (u, v) is larger than the variance of the number of pseudo-observations in the same square.

Nonparametric Estimation of the Measure of Functional Dependence

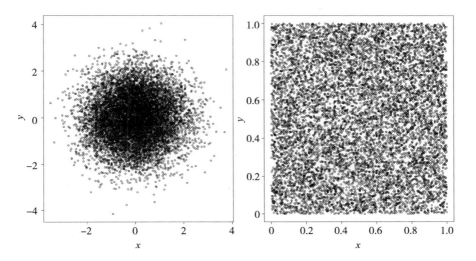

Figure 5-1 Scatter plots of random sample (n=10000) from standard normal distributions with $\rho=0$ (left) and its pseudo observations (right)

5.1.2 Kernel estimation through copula density functions

Kernel estimation $(X_1, Y_1), (X_2, Y_2), \cdots (X_n, Y_n)$ is a i.i.d. sample from $H_{X,Y}(x, y)$ with unknown marginals. The corresponding copula is estimated via pseudo-observations $(\hat{F}(X_i), \hat{G}Y_i)$, where $\hat{F}$ and $\hat{G}$ are the empiri-cal distribution functions

$$\hat{F}(x) = \frac{1}{n}\sum_{i=1}^{n} \mathbb{1}(X_i \leqslant x) \quad \text{and} \quad \hat{G}(y) = \frac{1}{n}\sum_{i=1}^{n} \mathbb{1}(Y_i \leqslant y)$$

Then, the estimator of the copula density function c is

$$\hat{c}(u, v) = \frac{1}{nh^2}\sum_{i=1}^{n} K\left(\frac{u - \hat{F}(X_i)}{h}, \frac{v - \hat{G}(Y_i)}{h}\right)$$

for bivariate kernel $K: R^2 \mapsto R$. It is asymptotically normal at every point $(u, v) \in (0, 1)^2$, i.e.

$$\frac{\hat{c}(u,v) - E(\hat{c}(u,v))}{\sqrt{Var(\hat{c}(u,v))}} \xrightarrow{D} N(0,1)$$

基于Copula的相关性测度
Measures of Association and Dependence Through Copulas

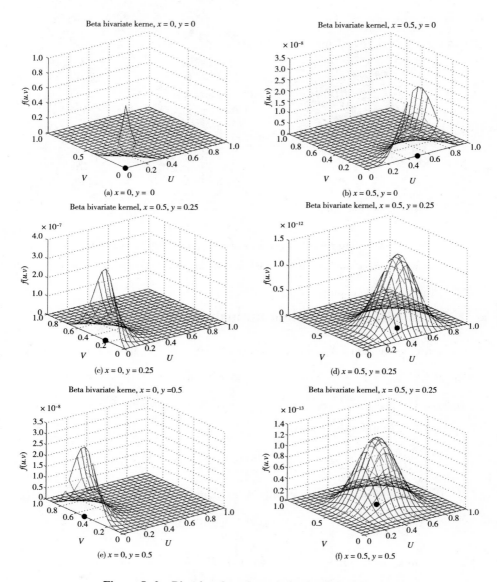

Figure 5-2　Bivariate beta kernels for bandwith $h=0.1$

Nonparametric Estimation of the Measure of Functional Dependence

Since the domain of c is $[0, 1]^2$, we choose Beta kernel. The idea of Beta kernel was introduced by Chen. It is one way to remove the boundary bias of usual kernel estimators. Let $X_1, \cdots, X_n$ be a random sample from a univariate distribution with density f having support $[0, 1]$, then the Beta kernel estimator of f is given by

$$\hat{f}_h(x) = \frac{1}{n}\sum_{i=1}^{n} K(X_i, \frac{x}{h}+1, \frac{1-x}{h}+1)$$

where $K(\alpha, \beta)$ is the density of the beta distribution with parameters α and β, i. e.,

$$K(x, \alpha, \beta) = \frac{x^{\alpha-1}(1-x)^{\beta-1}}{B(\alpha, \beta)}, \quad x \in I$$

Following the same idea, the beta-kernel based estimator of the copula density at point (u, v) is

$$\hat{c}_h(u, v) = \frac{1}{n}\sum_{i=1}^{n} K(u, \frac{X_i}{h}+1, \frac{1-X_i}{h}+1) K(v, \frac{Y_i}{h}+1, \frac{1-Y_i}{h}+1) \quad (5\text{-}2)$$

Assume that the copula density c is twice differentiable on $[0, 1]\times[0, 1]$. Let $(u, v) \in [0, 1]\times[0, 1]$. The bias of $\hat{c}$ is of order b, i. e.

$$E(\hat{c}(u, v)) = c(u, v) + \varrho(u, v) \cdot b + o(b)$$

where the bias $\varrho(u, v)$ is

$$\varrho(u, v) = (1-2u)\partial_1 c(u, v) + (1-2v)\partial_2 c(u, v) + \frac{1}{2}[u(1-u)\partial_{1,1}c(u, v) + v(1-v)\partial_{2,2}c(u, v)]$$

The variance of $\hat{c}(u, v)$ is in corners, e.g. $(0, 0)$

$$Var(\hat{c}(0, 0)) = \frac{1}{nh^2}[c(0, 0) + o(n^{-1})]$$

in the interior of borders, e.g. $u = 0$ and $v \in (0, 1)$

$$Var(\hat{c}(0, v)) = \frac{1}{2nh^{3/2}\sqrt{\pi v(1-v)}}[c(0, v) + o(n^{-1})]$$

and in the interior, $(u, v) \in (0, 1) \times (0, 1)$

$$Var(\hat{c}(u,v)) = \frac{1}{4nh\pi\sqrt{v(1-v)u(1-u)}}[c(u,v) + o(n^{-1})]$$

The optimal bandwidth h can be deduced by maximizing asymptotic mean squared error,

$$h^* = \left(\frac{1}{16\pi n \varrho(u,v)^2} \frac{1}{\sqrt{v(1-v)u(1-u)}}\right)^{1/3}$$

$\hat{c}_h(u,v)$ is asymptotically normally distributed :

$$\sqrt{nh^{k'}}[\hat{c}_h(u,v) - c(u,v)] \xrightarrow{w} N(0, \sigma(u,v)^2)$$

as $nh^{k'} \to \infty$ and $h \to 0$, where k' depends on the location, and $\sigma(u,v)^2$ is proportional to $c(u,v)$.

5.1.3 Asymptotic behavior of the estimator of functional dependence

We know that the "standardized" form of

$$\rho(Y \mid X) = \int_0^1 \int_0^1 \left(\frac{\partial C}{\partial u}\right)^2 du dv$$

and

$$\rho(X \mid Y) = \int_0^1 \int_0^1 \left(\frac{\partial C}{\partial v}\right)^2 du dv$$

are the measures of functional dependence. We will provide a kernel estimator of $\rho(Y \mid X)$ and discuss its asymptotic behavior. The asymptotic behavior of the estimator of $\rho(X \mid Y)$ can be obtained similarly. Since

$$\rho(Y \mid X) = \int_0^1 \int_0^1 \left(\frac{\partial C}{\partial u}\right)^2 du dv = \int_0^1 \int_0^1 \left(\int_0^v c(u,y) dy\right)^2 du dv$$

apply the beta kernel estimator of $c(u, v)$ in (5-2), we have the estimator of $\rho(Y \mid X)$:

Nonparametric Estimation of the Measure of Functional Dependence

$$\hat{\rho}(Y\mid X)=\int_0^1\int_0^1\left(\int_0^v\hat{c}(u,y)dy\right)^2 dudv \qquad (5\text{-}3)$$

To prove this estimator is asymptotically convergent, we need to use delta method.

First, we define ϕ_1 to be

$$\phi_1:c(u,v)\rightarrow\int_0^1\int_0^1\left(\int_0^v c(u,y)dy\right)^2 dudv$$

According to the Delta method,

$$\sqrt{nh^k}[\phi_1(\hat{c}_h(u,v))-\phi_1(c(u,v))]\xrightarrow{D}\phi_1'(N(0,\sigma(u,v)^2))$$

the Hadamard derivative of ϕ_1 is:

$$\frac{\phi_1(C+th)-\phi_1(C)}{t}$$

$$=\frac{1}{t}\int_0^1\int_0^1\left[\left(\int_0^v(C(u,y)+th(u,y))dy\right)^2-\left(\int_0^v(C(u,y))dy\right)^2\right]dudv$$

$$=\int_0^1\int_0^1\frac{1}{t}\left(\int_0^v(C(u,y)+th(u,y))dy+\int_0^v C(u,y)dy\right)$$

$$\left(\int_0^v(C(u,y)+th(u,y))dy-\int_0^v C(u,y)dy\right)dudv \qquad (5\text{-}4)$$

$$=\int_0^1\int_0^1\left(\int_0^v(C(u,y)+th(u,y))dy+\int_0^v C(u,y)dy\right)$$

$$\left(\int_0^v(h(u,y))dy\right)dudv$$

So the Hadamard derivative of ϕ_1 at C is:

$$\phi_1'(h)=\int_0^1\int_0^1\left(2\int_0^v C(u,y)dy\int_0^v h(u,y)dy\right)dudv$$

Therefore

$$\phi_1'(N(0,\sigma(u,v)^2))=\int_0^1\int_0^1\left(2\int_0^v C(u,y)dy\int_0^v N(0,\sigma(u,y)^2)dy\right)dudv$$

Since $\phi_1'(c(u,v))$ is a continuous and linear map, according to the delta method, $\hat{\rho}(Y\mid X)$ asymptotically converges to

$$N\left(0,\sigma(u,v)^2\left[\int_0^1\int_0^1\left(2\int_0^v C(u,y)dy\int_0^v N(0,\sigma(u,y)^2)dy\right)dudv\right]^2\right)$$

基于Copula的相关性测度
Measures of Association and Dependence Through Copulas

5.2 Nonparametric estimation of the measure of MCD via copula

Here we will discuss another possible way of estimating the measures of MCD or functional dependence. From the definition of the norm of a copula in (Definition 4.4), we obtain the estimator:

$$\hat{\rho} = \|\hat{C}\| = \int_0^1 \int_0^1 \left[\left(\frac{\partial \hat{C}}{\partial u}\right)^2 + \left(\frac{\partial \hat{C}}{\partial v}\right)^2\right] du dv$$

To discuss asymptotic distribution of $\|\hat{C}\|$, we need to have the estimator of copula first, which has been discussed in several papers. Instead of the empirical estimator based on pseudo-observations mentioned above, we mention two more estimators.

(1) Smoothed empirical copula:

Fermanian proposed a smoothed empirical distribution function $\hat{\mathbb{H}}_n(x, y)$,

$$\hat{\mathbb{H}}_n(x, y) = \frac{1}{n} \sum_{i=1}^n K_n(x - X_i, y - Y_i)$$

Here $K_n(x, y) = K(a_n^{-1}x, a_n^{-1}y)$, and

$$K(x, y) = \int_{-\infty}^x \int_{-\infty}^y k(x, y) du dv$$

for some bivariate kernel function $k: \mathbb{R}^2 \mapsto \mathbb{R}$, with $\int k(x, y) dx dy = 1$, and a sequence of bandwidth $a_n \downarrow 0$ as $n \to \infty$. Let $\hat{\mathbb{F}}_n(x)$ and $\mathbb{G}_n(y)$ be its associated marginal distributions.

Define smoothed empirical copula function $\hat{C}_n$ by

$$\hat{\mathbb{C}}_n = \hat{\mathbb{H}}_n(\hat{\mathbb{F}}_n^-(u), \hat{\mathbb{G}}_n^-(v))$$

and the empirical copula process

$$\hat{\mathbb{Z}}_n(u, v) = \sqrt{n}(\hat{\mathbb{C}}_n - C)(u, v)$$

Nonparametric Estimation of the Measure of Functional Dependence

According to Theorem 10 of Reference, the smoothed empirical copula process $\{\hat{\mathbb{Z}}_n(u,v), 0 \leq u, v \leq 1\}$ converges weakly to the Gaussian process $\{\mathbb{G}_c(u,v), 0 \leq u, v \leq 1\}$ in $\ell^\infty(\mathbf{I}^2)$.

The limiting Gaussian process can be written as
$$\mathbb{G}_c(u,v) = \mathbb{B}_c(u,v) - \partial_1 C(u,v)\mathbb{B}_c(u,1) - \partial_2 C(u,v)\mathbb{B}_c(1,v)$$
where $\mathbb{B}_c(u,v)$ is a Brownian bridge on $\mathbf{I}^2$ with covariance function
$$E[\mathbb{B}_c(u,v), \mathbb{B}_c(u',v')] = C(u \wedge u', v \wedge v') - C(u,v)C(u',v')$$
for each $0 \leq u, u', v, v' \leq 1$.

(2) Local linear kernel estimator: Chen and Huang constructed an estimator in the following way.
$$\hat{F}_n(x) = \frac{1}{n}\sum_{i=1}^n K\left(\frac{x-X_i}{b_{n1}}\right), \quad \hat{G}_n(y) = \frac{1}{n}\sum_{i=1}^n K\left(\frac{y-Y_i}{b_{n2}}\right)$$
with K the integral of a symmetric bounded kernel function k supported on $[-1,1]$, i.e. $K(x) = \int_{-\infty}^x k(t)dt$. Then the pseudo-observations $\hat{U}_i = \hat{F}_n(x)$ and $\hat{V}_i = \hat{G}_n(y)$ are used to estimate the joint distribution function. To prevent the boundary bias, they use a local linear version of k,
$$k_{u,h}(x) = \frac{k(x)\{a_2(u,h) - a_1(u,h)x\}}{a_0(u,h)a_2(u,h) - a_1^2(u,h)} \mathbf{1}\left\{\frac{u-1}{h} < x < \frac{x}{h}\right\}$$
where
$$a_l(u,h) = \int_{(u-1)/h}^{u/h} t^l k(t)dt$$
for $l = 0, 1, 2$. Finally, the local linear type estimator of the copula is given by
$$\hat{C}_n^{LL}(u,v) = \frac{1}{n}\sum_{i=1}^n K_{u,h_n}\left(\frac{u-\hat{U}_i}{h_n}\right) K_{v,h_n}\left(\frac{v-\hat{V}_i}{h_n}\right)$$

For the weak convergence results we need the following results,
$$\sup_x |\hat{F}_n(x) - F_n(x)| = o_p\left(\frac{1}{\sqrt{n}}\right) \tag{5-5}$$
$$\sup_y |\hat{G}_n(y) - G_n(y)| = o_p\left(\frac{1}{\sqrt{n}}\right)$$

which further implies the weak convergence of the processes $\sqrt{n}(\hat{F}_n - F)$ and $\sqrt{n}$

($\hat{G}_n - G$) to some Brownian bridges.

Like empirical copula estimator, this Local Linear Kernel estimator also converges to Gaussian process.

Theorem 5.1 Suppose that H has continuous marginal distribution functions and that the underlying copula function C has bounded second order partial derivatives on I^2. If $hn = O(n^{-1/3})$ and (5-5) is satisfied, then the copula process $\mathbb{C}_n^{LL} = \sqrt{n}(\hat{C}_n^{LL}(u,v) - C(u,v))$ converges weakly to the Gaussian process $\mathbb{G}_C$.

Define $\phi : \mathfrak{C} \mapsto \mathbb{R}$ by $\phi(C) = \|C\|^2$. One important step in discussing the asymptotic behavior of $\hat{\rho}$ is to see whether ϕ is Hadamard differentiable. Next we will show that it is indeed differentiable. Let $\mathbb{D}_2^1(I^2)$ be the Sobolev space, $D_1, D_2 \in \mathbb{D}_2^1(I^2)$, define inner product

$$\langle D_1, D_2 \rangle = \int_{I^2} \nabla D_1 \cdot \nabla D_2 d\lambda$$

The Sobolev norm induced by the inner product in $\mathbb{D}_2^1(I^2)$ is: for $D \in \mathbb{D}_2^1(I^2)$,

$$\|D\|^2 = \langle D, D \rangle = \int_{I^2} \left[\left(\frac{\partial D}{\partial u}\right)^2 + \left(\frac{\partial D}{\partial v}\right)^2 \right] dudv$$

Let $\mathfrak{C} \subset \mathbb{D}_2^1(I^2)$ be the copula space and $C \in \mathfrak{C}$ is a copula.

The derivative of ϕ at C along D is:

$$\lim_{t \downarrow 0} \frac{\phi(C+tD) - \phi(C)}{t}$$

$$= \lim_{t \downarrow 0} \frac{\int_{\mathbb{R}^2} \left[\left(\frac{\partial(C+tD)}{\partial u}\right)^2 + \left(\frac{\partial(C+tD)}{\partial v}\right)^2 \right] dudv - \int_{\mathbb{R}^2} \left[\left(\frac{\partial C}{\partial u}\right)^2 + \left(\frac{\partial C}{\partial v}\right)^2 \right] dudv}{t}$$

$$= 2 \int_{\mathbb{R}^2} \left[\frac{\partial C}{\partial u} \frac{\partial D}{\partial u} + \frac{\partial C}{\partial v} \frac{\partial D}{\partial v} \right] dudv$$

$$= 2\langle C, D \rangle$$

Nonparametric Estimation of the Measure of Functional Dependence

Since the above equalities hold for all $D \in \mathbb{D}_2^1(\mathbf{I}^2)$, ϕ is differentiable at C. Then with the fact that inner product is linear and continuous, we can conclude that ϕ is *Gâteaux* differentiable at C. Next, check that ϕ is Hadamard differentiable.

$$\lim_{n \to \infty} \frac{1}{t_n}(\phi(C + t_n D_n) - \phi(C))$$

$$= \lim_{n \to \infty} \frac{1}{t_n}(\langle C + t_n D_n, C + t_n D_n \rangle - \langle C, C \rangle)$$

$$= \lim_{n \to \infty} \frac{1}{t_n}(\langle C, C \rangle + 2\langle t_n D_n, C \rangle + \langle t_n D_n, t_n D_n \rangle - \langle C, C \rangle)$$

$$= \lim_{n \to \infty} 2\langle D_n, C \rangle$$

$$= 2\langle D, C \rangle$$

for all $t_n \to 0$, $D_n \in \mathbb{D}_2^1(\mathbf{I}^2)$, $C + t_n D_n \in \mathbb{D}_2^1(\mathbf{I}^2)$ and $D_n \to D$. Therefore, ϕ is Hadamard differentiable.

5.3 Simulation results

We report results from simulation studies. Three copulas are considered in the simulation study, which are respectively

$$C_\rho(u, v) = \int_{-\infty}^{\Phi^{-1}(u)} \int_{-\infty}^{\Phi^{-1}(v)} \frac{1}{2\pi\sqrt{1-\rho^2}} \exp\left\{-\frac{t^2 + s^2 - 2\rho ts}{2(1-\rho^2)}\right\} dtds$$

the Gaussian copula family with correlation coefficient $\rho = 0.5$;

$$C_\theta(u, v) = \exp\left(-[(-\ln u)^\theta + (-\ln v)^\theta]^{1/\theta}\right)$$

the Gumbel copula family with $\theta = 2$;

$$C_\theta(u, v) = uv + \theta u(1-u)v(1-v)$$

the Farlie-Gumbel-Morgenstern (FGM) family with $\theta = 1$.

Figure 5-3, Figure 5-4 and Figure 5-5 shows the beta kernel with bandwidth $h = 0.05$ estimations of the corresponding copula density functions. From

基于Copula的相关性测度
Measures of Association and Dependence Through Copulas

those figures we can see that the boundary biases are small, and as the sample size increase, the estimations are getting better. Figure 5-6, Figure 5-7 and Figure 5-8 are the density functions estimations using beta kernel with bandwidth $h = 0.01$. Since the bandwidth is too small, the surfaces are undersmoothed. From Table 5-4, Table 5-5 and Table 5-6, we can see that the estimation results for small sample size $n = 50$ is almost unacceptable. But for $n = 1000$, the results are generally better than the estimations with bandwidth $h = 0.05$.

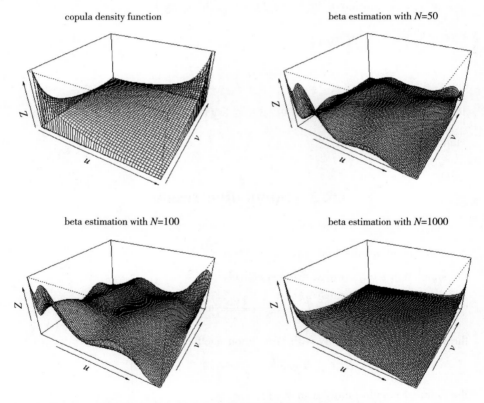

Figure 5-3   beta kernel with bandwidth $h = 0.05$ estimation of bivariate normal copula density function with $r = 0.5$

Nonparametric Estimation of the Measure of Functional Dependence

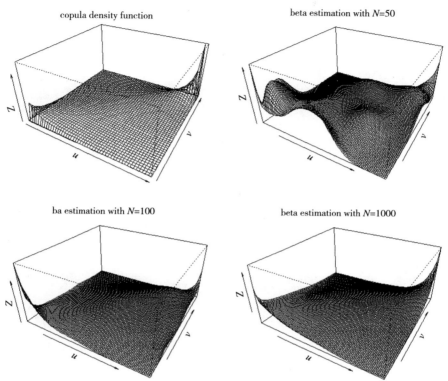

Figure 5-4 beta kernel with bandwidth $h=0.05$ estimation of Gumbel copula density function with $\theta=2$

Table 5-1 beta estimation of bivariate normal copula with $r=0.5$

Sample size	$\rho(Y\mid X)$	$\hat{\rho}(Y\mid X)$	$\rho(X\mid Y)$	$\hat{\rho}(Y\mid X)$	$\rho(X,Y)$	$\hat{\rho}(X,Y)$
$n=50$	0.3564	0.3796	0.3564	0.3959	0.7128	0.7756
$n=100$	0.3564	0.3414	0.3564	0.3502	0.7128	0.6916
$n=1000$	0.3564	0.3221	0.3564	0.3305	0.7128	0.6526

Table 5-2 beta estimation of Gumbel copula with $\theta=2$

Sample size	$\rho(Y\mid X)$	$\hat{\rho}(Y\mid X)$	$\rho(X\mid Y)$	$\hat{\rho}(Y\mid X)$	$\rho(X,Y)$	$\hat{\rho}(X,Y)$
$n=50$	0.3858	0.4178	0.3858	0.4350	0.7715	0.8528
$n=100$	0.3858	0.3455	0.3858	0.3523	0.7715	0.6978
$n=1000$	0.3858	0.3401	0.3858	0.3403	0.7715	0.6805

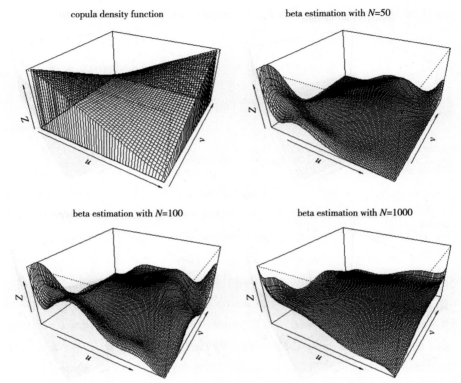

Figure 5-5 beta kernel with bandwidth $h=0.05$ estimation of FGM copula density function with $\theta=1$

Table 5-3 beta estimation of FGM copula with $\theta=1$

Sample size	$\rho(Y\mid X)$	$\hat{\rho}(Y\mid X)$	$\rho(X\mid Y)$	$\hat{\rho}(Y\mid X)$	$\rho(X,Y)$	$\hat{\rho}(X,Y)$
$n=50$	0.3440	0.4455	0.3440	0.4831	0.6880	0.9287
$n=100$	0.3440	0.3395	0.3440	0.3734	0.6880	0.7194
$n=1000$	0.3440	0.3140	0.3440	0.3143	0.6880	0.6282

Table 5-4 beta estimation of Gumbel copula with $\theta=2$

Sample size	$\rho(Y\mid X)$	$\hat{\rho}(Y\mid X)$	$\rho(X\mid Y)$	$\hat{\rho}(Y\mid X)$	$\rho(X,Y)$	$\hat{\rho}(X,Y)$
$n=50$	0.3858	0.5096	0.3858	0.5206	0.7715	1.0301
$n=100$	0.3858	0.4203	0.3858	0.4282	0.7715	0.8485
$n=1000$	0.3858	0.3860	0.3858	0.3864	0.7715	0.7724

Nonparametric Estimation of the Measure of Functional Dependence

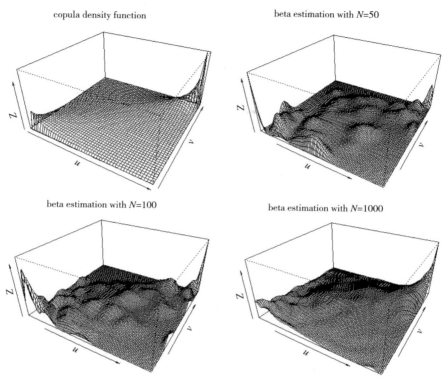

Figure 5-6 beta kernel with bandwidth $h=0.01$ estimation of Gumbel copula density function with $\theta=2$

Table 5-5 beta estimation of bivariate normal copula with $r=0.5$

Sample size	$\rho(Y\mid X)$	$\hat{\rho}(Y\mid X)$	$\rho(X\mid Y)$	$\hat{\rho}(Y\mid X)$	$\rho(X,Y)$	$\hat{\rho}(X,Y)$
$n=50$	0.3564	0.4886	0.3564	0.4935	0.7128	0.9822
$n=100$	0.3564	0.4183	0.3564	0.4095	0.7128	0.8279
$n=1000$	0.3564	0.3701	0.3564	0.3791	0.7128	0.7492

Table 5-6 beta estimation of FGM copula with $\theta=1$

Sample size	$\rho(Y\mid X)$	$\hat{\rho}(Y\mid X)$	$\rho(X\mid Y)$	$\hat{\rho}(Y\mid X)$	$\rho(X,Y)$	$\hat{\rho}(X,Y)$
$n=50$	0.3440	0.5200	0.3440	0.6180	0.6880	1.1381
$n=100$	0.3440	0.3932	0.3440	0.4427	0.6880	0.8360
$n=1000$	0.3440	0.3534	0.3440	0.3549	0.6880	0.7084

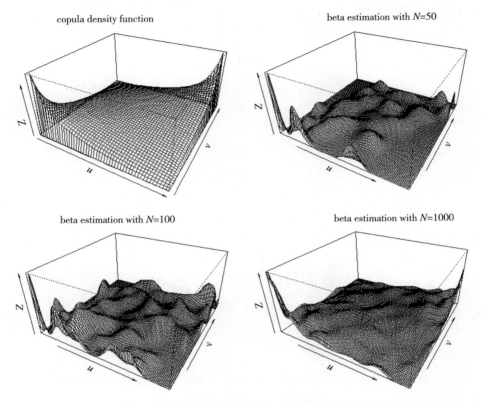

Figure 5-7　beta kernel with bandwidth $h=0.01$ estimation of bivariate normal copula density function with $r=0.5$

Nonparametric Estimation of the Measure of Functional Dependence

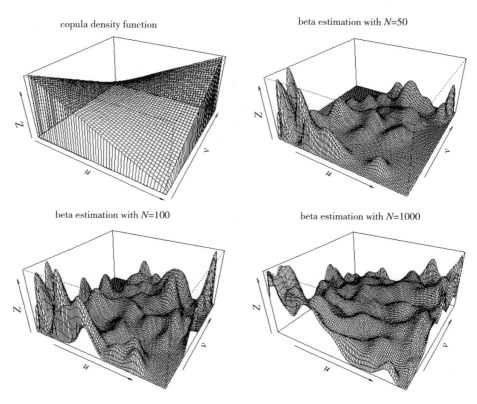

Figure 5-8 beta kernel with bandwidth $h=0.01$ estimation of FGM copula density function with $\theta=1$

6

Implementation and Simulations

6.1 Choosing the evaluation grid

In copula density estimation, Nagler suggested to use a grid that is equally spaced after a transformation by the inverse Gaussian cdf, which is shown in Figure 6-1. Our simulation results below show that evaluating copula density at a set of grid points in similar pattern will improve the accuracy of estimators of MCD. To compare the impact of the choice of grid, we considered two copula families, the Gaussian copula with parameters 0, 0.1, 0.2, 0.5, 0.8, 0.9 and the Gumbel copula with parameters 1, 10/9, 10/7, 10/3, 5, 10. Two samples of sizes 200 and 1000 were taken from each copula respectively. First, copula densities are estimated from each sample based on the KDE copula package. Then, the estimated copula density was evaluated on two sets of grid points: the usual grid with equally spaced points and a normalized grid. From the discretized copula density we calculate the estimate of MFD. Figure 6-2 and Figure 6-3 show the mean absolute error (MAE) of estimators with sample size 200 and 1000 under 500 replication. In each case, we find the MAE of estimators based on equally spaced grid, labeled "equal", are significantly higher than the MAE of the same estimators based on the transformed grid, labeled "norm".

Implementation and Simulations

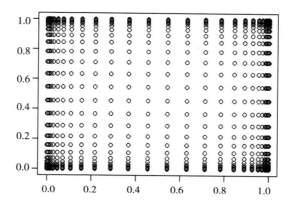

Figure 6-1 A grid which is equally spaced after inverse Gaussian cdf transformation

6.2 Simulation

In this section, we explore the finite sample performance of the proposed estimators using the mean squared error (MSE). To put all estimators on the same scale, we standardize MFD. In other words, we use ρ_1, ρ_2 and ρ. The corresponding estimators will be denoted by $\hat{\rho}_1 (Y \mid X)$, $\hat{\rho}_2 (X \mid Y)$ and $\hat{\rho} (X, Y)$. For twodimensional density estimation, using cross-validation to choose the bandwidth is computationally expensive. Therefore, in all simulations, a rule-of-thumb bandwidth is used. More precisely, the bandwidth is selected based on the asymptotic mean integrated squared error (AMISE) -optimality with respect to the Frank copula. For further details on bandwidth selection in this context, we refer to reference. In all simulations reported here, the integration is calculated over a grid of 30 × 30 points. For the choice of grid, we adopt the method in reference. That is, we apply the Gaussian cumulative distribution function to equally spaced 30 knots on a line segment $[-3, 3]$. The final two-dimensional grid is shown in Figure 6-1. This choice takes into account the fact that copula densities usually have high fluctuation on the boundary and corners. Putting more evaluating points on those regions will reduce approximation errors.

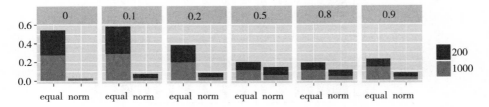

Figure 6-2 MAE of estimators of MFD for samples drawn from normal copula

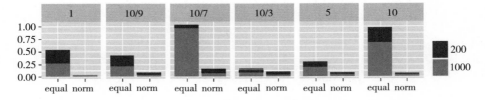

Figure 6-3 MAE of estimators of MFD for samples drawn from Gumbel copula

In Tables 6-1 and Tables 6-2, we present the simulated MSE of the estimators of ρ_1 and ρ_2 for samples size 50, 100 and 200 for Gaussian copula. These results are based on 1000 replications. Both copulas are generated by R package "copula". We find that our estimators have reasonable precision in all cases. As sample size increases, MSE is getting smaller. And there is no significant difference in MSE for different θ values, which indicates our estimator is stable for the choice of θ's.

Implementation and Simulations

Table 6-1 Simulated MSE of the estimates when the underlying copula is Gaussian copula with correlation θ

		$n=50$	$n=100$	$n=200$
$\theta=0$	$\hat{\rho}_1(Y\mid X)$	2.3×10^{-3}	1.5×10^{-3}	8.0×10^{-4}
	$\hat{\rho}_2(X\mid Y)$	2.3×10^{-3}	1.5×10^{-3}	7.3×10^{-3}
$\theta=0.3$	$\hat{\rho}_1(Y\mid X)$	6.1×10^{-3}	3.8×10^{-3}	2.2×10^{-3}
	$\hat{\rho}_2(X\mid Y)$	6.2×10^{-3}	3.8×10^{-3}	2.1×10^{-3}
$\theta=0.6$	$\hat{\rho}_1(Y\mid X)$	8.8×10^{-3}	4.2×10^{-3}	2.1×10^{-3}
	$\hat{\rho}_2(X\mid Y)$	8.9×10^{-3}	4.1×10^{-3}	2.1×10^{-3}
$\theta=0.9$	$\hat{\rho}_1(Y\mid X)$	2.5×10^{-3}	1.1×10^{-3}	4.9×10^{-4}
	$\hat{\rho}_2(X\mid Y)$	2.3×10^{-3}	1.1×10^{-3}	5.0×10^{-4}

Table 6-2 Simulated MSE of the estimates when the underlying copula is Clayton copula with parameter θ

		$n=50$	$n=100$	$n=200$
$\theta=0.2$	$\hat{\rho}_1(Y\mid X)$	3.7×10^{-3}	2.3×10^{-3}	1.8×10^{-3}
	$\hat{\rho}_2(X\mid Y)$	3.7×10^{-3}	2.3×10^{-3}	1.8×10^{-3}
$\theta=0.5$	$\hat{\rho}_1(Y\mid X)$	7.2×10^{-3}	4.0×10^{-3}	2.3×10^{-3}
	$\hat{\rho}_2(X\mid Y)$	7.2×10^{-3}	4.1×10^{-3}	2.3×10^{-3}
$\theta=1$	$\hat{\rho}_1(Y\mid X)$	9.3×10^{-3}	4.9×10^{-3}	2.6×10^{-3}
	$\hat{\rho}_2(X\mid Y)$	9.3×10^{-3}	4.9×10^{-3}	2.5×10^{-3}
$\theta=2$	$\hat{\rho}_1(Y\mid X)$	7.5×10^{-3}	3.5×10^{-3}	1.8×10^{-3}
	$\hat{\rho}_2(X\mid Y)$	7.4×10^{-3}	3.5×10^{-3}	1.9×10^{-3}
$\theta=5$	$\hat{\rho}_1(Y\mid X)$	3.1×10^{-3}	1.2×10^{-3}	6.1×10^{-4}
	$\hat{\rho}_2(X\mid Y)$	3.1×10^{-3}	1.2×10^{-3}	6.0×10^{-4}

基于Copula的相关性测度
Measures of Association and Dependence Through Copulas

6.3 Comparison of measures

In the second part of simulation, we will compare the performance of MFDs with other measures of dependence, e.g., linear correlation coefficient r, Spearman's ρ and Kendall's τ under several different types of relationships. We choose three different dependence structures: elliptical distributions, monotonic dependence and regressional dependence, represented by normal copula, cubic function and quadratic function respectively.

The first example is a quadratic function. 500 data are generated from the following model,

$$Y = X^2 + \varepsilon \qquad (6-1)$$

where $\varepsilon \sim N(0, \sigma)$ and $\sigma = 1, 5$ and 10. To obtain the copula data, we apply the empirical marginal distributions to the data. Then beta kernel estimation is applied to get the estimations of ρ_1 and ρ_2.

Obviously, neither Spearman's ρ nor Kendall's τ are suitable for this situation. The simulation results in Table 6-3 also showed that they are almost 0 in all cases. $\hat{\rho}(X, Y)$, on the other hand, is much higher than both Spearman's ρ and Kendall's τ, especially for small σ. This indicates that the type of dependence is functional, not monotonic. And the magnitude of $\hat{\rho}(X, Y)$ tells that the strength of dependence is getting weaker as σ increases. In comparison of $\hat{\rho}_1(Y|X)$ and $\hat{\rho}_2(X|Y)$, we find that functional dependence is stronger in the Y to X direction than the other direction, since $\hat{\rho}_1(Y|X)$ is higher. Again, as σ increases, the strength of dependence in this direction is also getting weaker.

6 Implementation and Simulations

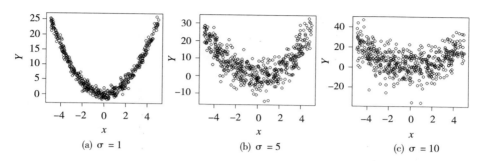

(a) $\sigma = 1$ (b) $\sigma = 5$ (c) $\sigma = 10$

Figure 6-4 Scatter plot of model $Y=X_2+\varepsilon$, sample size $N=500$

Table 6-3 Estimators based on a sample of size 500 when the underlying relationship is parabola

	$\hat{\rho}_1 (Y\mid X)$	$\hat{\rho}_2 (X\mid Y)$	$\hat{\rho} (X, Y)$	Spearman's ρ	Kendall's τ
$\sigma = 1$	0.51	0.25	0.40	−0.01	−0.02
$\sigma = 5$	0.41	0.20	0.32	−0.02	−0.02
$\sigma = 10$	0.27	0.13	0.21	−0.01	−0.01

To compare the performance of MFD with Kendall's τ and Spearman's ρ in monotonic dependence, 500 data are generated from the following model,

$$Y=X^3+\varepsilon \qquad (6-2)$$

where $\varepsilon \sim N(0, \sigma)$ and $\sigma=1, 5$ and 10. The scatter plot of model (6-2) is in Figure 6-5 and the simulation results are in Table 6-4. As shown in Table 6-4, the values of MFD has similar decreasing pattern as the other two measures when σ increases. Indeed, cubic function is one type of functional dependence, so MFDs are capable of measuring the strength of monotonic dependence. The values of $\hat{\rho}_1 (Y\mid X)$ and $\hat{\rho}_2 (X\mid Y)$, which measure the strength of functional dependence in two directions (Y to X and X to Y) separately, are close to each other, obviously this is because model (6-2) is symmetric.

Next, we take into account the Pearson's correlation coefficient. We take

基于Copula的相关性测度
Measures of Association and Dependence Through Copulas

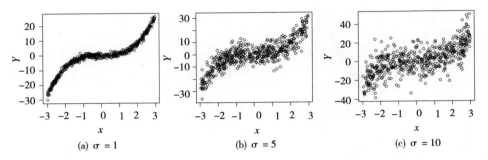

(a) σ = 1   (b) σ = 5   (c) σ = 10

Figure 6-5 Scatter plot of model $Y=X^3+\varepsilon$, sample size $N=500$

normal copulas as an example, which is with $r = 0.1$, 0.5 and 0.9. The scatter plots of Gaussian copulas are in Figure 6-6. Table 6-5 shows the simulation results. As expected, those measures show no significant difference in measuring dependence of elliptical distributions.

Table 6-4 Estimators based on a sample of size 500 when the underlying relationship is cubic

	$\hat{\rho}_1$ (Y|X)	$\hat{\rho}_2$ (X|Y)	$\hat{\rho}$ (X, Y)	Spearman's ρ	Kendall's τ
σ = 1	0.89	0.89	0.89	0.82	0.95
σ = 5	0.63	0.62	0.63	0.55	0.74
σ = 10	0.42	0.41	0.41	0.35	0.50

$$C_r(u,v) = \int_{-\infty}^{\Phi^{-1}(u)} \int_{-\infty}^{\Phi^{-1}(v)} \frac{1}{2\pi\sqrt{1-r^2}} \exp\left\{-\frac{t^2+s^2-2rts}{2(1-r^2)}\right\} dtds \quad (6-3)$$

The comparison of MFDs with other measures in model (6-1), model (6-2) and model (6-3) shows that MFDs have a good adaptability for different types of relationships.

Implementation and Simulations

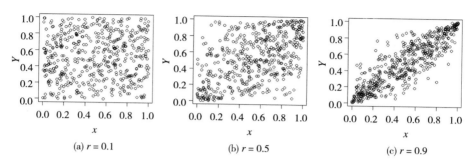

Figure 6-6 Normal copulas with parameter r

Table 6-5 Estimators based on a sample of size 500 when the underlying copula is Gaussian copula

	$\hat{\rho}_1\ (Y\mid X)$	$\hat{\rho}_2\ (X\mid Y)$	$\hat{\rho}\ (X,\ Y)$	Spearman's ρ	Kendall's τ
$r=0.1$	0.09	0.09	0.09	0.08	0.11
$r=0.5$	0.33	0.33	0.33	0.29	0.42
$r=0.9$	0.73	0.73	0.73	0.68	0.87

❼ Application

The Communities and Crime Data Set contains community crime rate of 1994 communities with 123 possibly related variables. We will use functional dependence measure as a criteria for variable selection to choose the variables which have the most impact on community crime rate. So the MFD between community crime rate and each of other variables were calculated. Table 7-1 shows 9 variables with highest functional dependence measures. As shown in Figure 5-2 of the selected variables, PctIlleg, racePctWhite showed clear nonlinear relations with the crime rate.

Table 7-1 Variables with highest scores in functional dependence measure

	variable	$\hat{\rho}_1 \ (Y|X)$	$\hat{\rho}_2 \ (X|Y)$	$\hat{\rho} \ (X, Y)$
1	PctKids2Par	0.56	0.57	0.57
2	PctIlleg	0.53	0.54	0.54
3	PctFam2Par	0.53	0.54	0.53
4	NumIlleg	0.52	0.53	0.53
5	PctTeen2Par	0.49	0.49	0.49
6	racePctWhite	0.49	0.49	0.49
7	FemalePctDiv	0.49	0.49	0.49
8	NumUnderPov	0.48	0.48	0.48
9	TotalPctDiv	0.48	0.48	0.48

Application

Table 7-2 Selected variables for the communities crime rate data

variable	Attribute
PctKids2Par	Percentage of kids in family housing with two parents
PctIlleg	Percentage of kids born to never married
PctFam2Par	Percentage of families (with kids) that are headed by two parents
NumIlleg	Number of kids born to never married
PctTeen2Par	Percent of kids age 12-17 in two parent households
racePctWhite	Percentage of population that is caucasian
FemalePctDiv	Percentage of females who are divorced
NumUnderPov	Number of people under the poverty level
TotalPctDiv	Percentage of population who are divorced

⑧ Discussion

The results showed that, compared with Spearman's rho or Kendall's tau, the measures of functional relationship can not only measure the strength of a relationship, but also indicate the direction of a possible functional relationship. We provide a novel method to estimate the measures of functional relationship. The simulation results showed that they have fairly good accuracy.

Although MFD can quantify the strength of functional dependence, it doesn't suggest any specific form of the function. So one possible application of this measure is in variable selection. We use MFD to filter out most of the unrelevant variables, then use parametric or nonparametric methods to construct a predicting model. In the community crime data example, we showed that MFD can detect nonlinear relationship, but as for how many variables should be retained, in other words, how to set up the threshold of MFD in variable selection is a question need to be discussed, and may involve some subjective opinions. After the desired number of variables are chosen, people may use either parametric or nonparametric methods to set up the model.

References

[1] Arthur Asuncion and David Newman. Uci machine learning repository. 2007.

[2] Javad Behboodian, Ali Dolati, and Manuel Ubeda-Flores. A multivariate version of gini's rank association coefficient [J]. Statistical Papers, 2007, 48 (2): 295-304.

[3] Arthur Charpentier, Jean-David Fermanian, and Olivier Scaillet. The estimation of copulas: Theory and practice [A]. Copulas: From theory to Application in Finance [M]. London: Risk Books, 2007: 35-60.

[4] Song Xi Chen. Beta kernel estimators for density functions [J]. Computational Statistics & Data Analysis, 1999, 31 (2): 131-145.

[5] Song Xi Chen and Tzee-Ming Huang. Nonparametric estimation of copula functions for dependence modelling [J]. Canadian Journal of Statistics, 2007, 35 (2): 265-282.

[6] Xiaohong Chen and Yanqin Fan. Estimation and model selection of semiparametric copula – based multivariate dynamic models under copula misspecification [J]. Journal of Econometrics, 2006, 135 (1): 125-154.

[7] W. F. Darsow, B. Nguyen, and E. T. Olsen. Copulas and markov processes [J]. Illinois Journal of Mathematics, 1992, 36 (4): 600-642.

[8] W. F. Darsow and E. T. Olsen. Norms for copulas [J]. International Journal of Mathematics and Mathematical Sciences, 1995, 18 (3): 417-436.

[9] E. de Amo, M. Díaz Carrillo, and J. Fernández-Sánchez. Characterization of all copulas associated with non – continuous random variables [J]. Fuzzy Sets and Systems, 2012 (191): 103-112.

[10] P. Deheuvels. La fonction de dépendance empirique et ses propriétés. Académie Royal de Belgique [J]. Bulletin de la Classe des Sciences, 1979, 65 (5): 274-292.

[11] Michel Denuit and Philippe Lambert. Constraints on concordance measures in bivariate discrete data [J]. Journal of Multivariate Analysis, 2005, 93 (1): 40-57.

[12] Michel Denuit and Olivier Scaillet. Nonparametric tests for positive quadrant dependence [J]. Journal of Financial Econometrics, 2004, 2 (3): 422-450.

[13] A. Dolati and M. Ubeda-Flores. On measures of multivariate concordance [J]. Journal of Probability and Statistical Science, 2006, 4 (2): 147-164.

[14] Paul Embrechts, Filip Lindskog, and Alexander McNeil. Modelling dependence with copulas and applications to risk management [J]. Handbook of Heavy Tailed Distributions in Finance, 2003, 8 (1): 329-384.

[15] Jean-David Fermanian, Dragan Radulovic, and Marten Wegkamp. Weak convergence of empirical copula processes [J]. Bernoulli, 2004, 10 (5): 847-860.

[16] Begña Fernández Fernández and José M Gonzánez-Barrios. Multidimensional dependency measures [J]. Journal of Multivariate Analysis, 2004, 89 (2): 351-370.

[17] C. Genest and J. Neslehova. A primer on copulas for count data [J]. Astin Bulletin, 2007, 37 (2): 475.

[18] Christian Genest, Kilani Ghoudi, and L-P Rivest. A semiparametric estimation procedure of dependence parameters in multivariate families of distributions [J]. Biometrika, 1995, 82 (3): 543-552.

[19] Christian Genest, Johanna Nešlehová, and Noomen Ben Ghorbal. Spearman's footrule and gini's gamma: A review with complements [J]. Journal of Nonparametric Statistics, 2010, 22 (8): 937-954.

[20] Christian Genest and Louis-Paul Rivest. Statistical inference proce-

References

dures for bivariate archimedean copulas [J]. Journal of the American Statistical Association, 1993, 88 (423): 1034-1043.

[21] Enzo Giacomini, Wolfgang Härdle, and Vladimir Spokoiny. Inhomogeneous dependence modeling with time - varying copulae [J]. Journal of Business & Economic Statistics, 2009, 27 (2).

[22] A. Janic-Wróblewska, WCM Kallenberg, and T. Ledwina. Detecting positive quadrant dependence and positive function dependence [J]. Insurance: Mathematics and Economics, 2004, 34 (3): 467-487.

[23] H. Joe. Multivariate concordance [J]. Journal of Multivariate Analysis, 1990, 35 (1): 12-30.

[24] Harry Joe. Multivariate models and multivariate dependence concepts [M]. CRC Press, 1997.

[25] Harry Joe. Asymptotic efficiency of the two-stage estimation method for copula-based models [J]. Journal of Multivariate Analysis, 2005, 94 (2): 401-419.

[26] Harry Joe and James J. Xu. The estimation method of inference functions for margins for multivariate models [J]. Technical Report, 1996.

[27] Gunky Kim, Mervyn J Silvapulle, and Paramsothy Silvapulle. Comparison of semiparametric and parametric methods for estimating copulas [J]. Computational Statistics & Data Analysis, 2007, 51 (6): 2836-2850.

[28] William H Kruskal. Ordinal measures of association [J]. Journal of the American Statistical Association, 1958, 53 (284): 814-861.

[29] Pranesh Kumar. Statistical dependence: Copula functions and mutual information based measures [J]. Journal of Statistics Applications & Probability: An International Journal, 2012, 1 (1): 1-14.

[30] HO Lancaster. Correlation and complete dependence of random variables [J]. The Annals of Mathematical Statistics, 1963: 1315-1321.

[31] HO Lancaster. Dependence, measures and indices of. Encyclopedia of Statistical Sciences, 1982.

[32] E. L. Lehmann. Some concepts of dependence [J]. The Annals of

Mathematical Statistics, 1966: 1137-1153.

[33] Yannick Malevergne and Didier Sornette. Extreme financial risks: From dependence to risk management [M]. Springer Science & Business Media, 2006.

[34] Alexander J. McNeil, Rüdiger Frey, and Paul Embrechts. Quantitative risk management: Concepts, techniques, and tools [M]. Princeton University Press, 2010.

[35] Mhamed Mesfioui and Abdelouahid Tajar. On the properties of some nonparametric concordance measures in the discrete case [J]. Nonparametric Statistics, 2005, 17 (5): 541-554.

[36] Thomas Nagler. Kdecopula: An r package for the kernel estimation of copula densities. arXiv preprint arXiv: 1603.04229, 2016.

[37] R. B. Nelsen. Nonparametric measures of multivariate association [J]. Lecture Notes-Monograph Series, 1996: 223-232.

[38] R. B. Nelsen. An introduction to copulas [M]. Springer, 2006.

[39] R. B. Nelsen. Copulas and association [A]. In Advances in Probability Distributions with Given Marginals [M]. Springer, 1991: 51-74.

[40] R. B. Nelsen. Concordance and copulas: A survey [A]. In Distributions with given marginals and statistical modelling [M]. Springer, 2002: 169-177.

[41] J. Nešlehová. On rank correlation measures for non-continuous random variables [J]. Journal of Multivariate Analysis, 2007, 98 (3): 544-567.

[42] Andrew John Patton. Applications of copula theory in financial econometrics [D]. PhD thesis, University of California, San Diego, 2002.

[43] Ludger Ruschendorf. Asymptotic distributions of multivariate rank order statistics [J]. The Annals of Statistics, 1976: 912-923.

[44] Cornelia Savu and Mark Trede. Goodness-of-fit tests for parametric families of archimedean copulas [J]. Quantitative Finance, 2008, 8 (2): 109-116.

[45] Marco Scarsini. On measures of concordance [J]. Stochastica: Revista de Matemática pura y Aplicada, 1984, 8 (3): 201-218.

References

[46] Friedrich Schmid and Rafael Schmidt. Bootstrapping spearmans multivariate rho [A]. // COMPSTAT, Proceedings in Computational Statistics [M]. Springer, 2006: 759-766.

[47] Friedrich Schmid and Rafael Schmidt. Multivariate extensions of spearman's rho and related statistics [J]. Statistics & Probability Letters, 2007, 77 (4): 407-416.

[48] Friedrich Schmid and Rafael Schmidt. Nonparametric inference on multivariate versions of blomqvists beta and related measures of tail dependence [J]. Metrika, 2007, 66 (3): 323-354.

[49] Berthold Schweizer and Edward F Wolff. On nonparametric measures of dependence for random variables [J]. The Annals of Statistics, 1981: 879-885.

[50] K. F. Siburg and P. A. Stoimenov. A measure of mutual complete dependence [J]. Metrika, 2010, 71 (2): 239-251.

[51] M Sklar. Fonctions de répartition `a n dimensions et leurs marges [J]. Université Paris 8, 1959.

[52] Santi Tasena and Sompong Dhompongsa. A measure of multivariate mutual complete dependence [J]. International Journal of Approximate Reasoning, 2013, 54 (6): 748-761.

[53] M. D. Taylor. Multivariate measures of concordance [J]. Annals of the Institute of Statistical Mathematics, 2007, 59 (4): 789-806.

[54] Manuel Ubeda-Flores. Multivariate versions of blomqvists beta and spearmans footrule [J]. Annals of the Institute of Statistical Mathematics, 2005, 57 (4): 781-788.

[55] Aad W. Van Der Vaart and Jon A. Wellner. Weak Convergence [M]. Springer, 1996.

[56] E. F. Wolff. N-dimensional measures of dependence [J]. Stochastica: Revista de Matemática Pura y Aplicada, 1980, 4 (3): 175-188.

Appendix

A List of Symbols

PQD: positively quadrant dependence

MCD: mutual complete dependence

$\mathcal{C}$: the set of discrete 2-subcopula

C: subcopula function

c: subcopula density function

H: joint distribution function

h: joint mass function

$\mathbb{1}$: indicator function

$\mapsto$: map

$\rightarrow$: convergence

$\xrightarrow{as}$: almost sure convergence

$\xrightarrow{w}$: weak convergence

$\xrightarrow{D}$: convergence in distribution

$\mathbb{R}$: Real number

Appendix

$\bar{\mathbb{R}}$: extended real numbers, $1[1-\infty, +\infty]$

$\mathbf{I} = [0,1]$

r: Pearson's correlation coefficient

ρ: Spearman's rank correlation

τ: Kendall's rank correlation

$\|C\|^2$: discrete norm of a copula C

$\partial_i f$: ith partial derivative of function f

C^+, C^-: Fréchet-Hoeffding boundaries

$E[X]$: expectation of X

Ran F: range of function F

$\Pi(u,v)$: independent copula

$\operatorname{supp}(X) = \{x \mid P(X = x) > 0\}$

$\mu_t(X, Y)$: the measure of MCD for discrete random variables X and Y

$\bar{\lambda}(X, Y)$: the measure of MCD for finite discrete random variables X and Y

$\mu_t(Y \mid X), \mu_t(X \mid Y)$: the measures of functional dependence between random variables X and Y

$\omega^2(Y \mid X)$: the expectation of weighted distance between the conditional distribution of Y given X and marginal distribution of Y

$\ell^\infty(T)$: set of all uniformly bounded real functions on T

B Calculation of the Measure of PQD

```r
# First example, copula is completely under U*V as -1 < rho < 0,
    completely above U*V as 0 < rho <= 1
#rho <- -0.7
#rho <- 1
#C <- function(u, v)
#{
# return (u * v + rho * u * v * (1 - u) * (1 - v))
#}

#Second example, copula is partly above u*v and partly below u*v.

C <- function(u, v)
{
    t <- 0.8
    if(u >= t & v >= t)
    {
        return (max(u + v - 1, t))
    }
    else
```

Appendix

```r
{
    return (min(u, v))
  }
}

N <- 1000
U <- seq(0, 1, 1 / N)
V <- seq(0, 1, 1 / N)
inte <- 0.0

for (i in 1:N)
{
  for(j in 1:N)
  {
    if(C(U[i], V[j]) < U[i] * V[j])
    {
      #t <- t + 1
      temp <- (C(U[i + 1], V[j]) - C(U[i], V[j])) / (U[i + 1] - U[i]) * V[j]
        + (C(U[i], V[j + 1]) - C(U[i], V[j])) / (V[j + 1] - V[j]) * U[i]
      #browser()
```

基于Copula的相关性测度
Measures of Association and Dependence Through Copulas

```
39        #print(temp * (1 / N) * (1 / N))
40        inte = inte + temp * (1 / N) * (1 / N)
41      }
42    }
43 }
44 PQD.measure = 1 - 3/2 * inte
45 print(PQD.measure)
```

PQD measure.R

C Beta Kernel Estimation

```
1  # functions for calculating:
2  # dens.rho1=\int \int [(\partial C)/(\partial u)]^2 du dv
3  # dens.rho2=\int \int [(\partial C)/(\partial v)]^2 du dv
4  # rho.sqr = dens.rho1 + dens.rho2
5  # dens.rho1 <- function(Z)
6  # {
7  #   # chat= c(x,y) density
8  #   delta1 <- 1/(nrow(Z))   # delta1 = delta x
9  #   delta2 <- 1/(ncol(Z)) # delta2 = delta y
10 #   chat1 <- Z
11 #   rho1.sqr <-0
12 #   for(i in 1:nrow(chat1)) {
13 #     for(j in 1:ncol(chat1)){
```

Appendix

```
14 #          density.inte <- sum((chat1[i,1:j-1]+chat1[i,2:j])*delta2/2)+
            chat1[i,1]*delta2/2
15 #           rho1.sqr <- rho1.sqr + density.inte^2*delta1*delta2
16 #       }
17 #    }
18 #    return(rho1.sqr)
19 # }
20
21 dens.rho1 <- function(Z)
22 {
23    # chat= c(x,y) density
24    delta1 <- 1/(nrow(Z))   # delta1 = delta x
25    delta2 <- 1/(ncol(Z))   # delta2 = delta y
26    chat1 <- Z
27    rho1.sqr <-0
28    for(i in 1:nrow(chat1)) {
29       for(j in 1:ncol(chat1)){
30          density.inte <- sum((chat1[i,1:j-1]+chat1[i,2:j])*delta2/2)
31          rho1.sqr <- rho1.sqr + density.inte^2*delta1*delta2
32       }
33    }
34    return(rho1.sqr)
35 }
36
37 # function- calculate \rho_2^2 (X is a function of Y) from the
    estimated copula density function
```

基于Copula的相关性测度
Measures of Association and Dependence Through Copulas

```r
38  dens.rho2 <- function(Z)
39  {
40    # chat= c(x,y) density
41    delta1 <- 1/(nrow(Z)) # delta1 = delta x
42    delta2 <- 1/(ncol(Z)) # delta2 = delta y
43    chat2 <- Z
44    rho2.sqr <-0
45    for(i in 1:nrow(chat2)) {
46      for(j in 1:ncol(chat2)){
47        density.inte <- sum((chat2[1:i-1,j]+chat2[2:i,j])*delta1/2) +
              chat2[1,j]*delta1/2
48        rho2.sqr <- rho2.sqr + density.inte^2*delta1*delta2
49      }
50    }
51    return(rho2.sqr)
52  }
53  # calculate rho^2 (the norm of C) from the copula density estimator
54  dens.rho <- function(chat)
55  {
56    rho.sqr <- dens.rho1(chat) + dens.rho2(chat)
57    return(rho.sqr)
58  }
59
60  #This function evaluate the density of a given copula at a P*q matrix
61  copula.density = function (p,q,copula) {
62    s <- seq(1/p, len=(p-1), by=1/p)
```

Appendix

```r
t <- seq(1/q, len=(q-1), by=1/q)
mat <- matrix(0, nrow = p-1, ncol = q-1)
for (i in 1:(p-1)) {
  a <- s[i]
  for (j in 1:(q-1)) {
    b <- t[j]
    mat[i,j] <- dCopula(c(a, b), copula)
  } }
  return(data.matrix(mat))
}

#beta kernel copula estimation
library(copula)
#(u,v) are sample points, bx,by are bandwith of X,Y,
# p number of grids in each directions,
beta.kernel.copula.surface = function (u,v,bx,by,p) {
  s <- seq(1/p, len=(p-1), by=1/p)
  mat <- matrix(0, nrow = p-1, ncol = p-1)
  for (i in 1:(p-1)) {
    a <- s[i]
    for (j in 1:(p-1)) {
      b <- s[j]
      mat[i,j] = sum(dbeta(a,u/bx,(1-u)/bx) *
```

基于Copula的相关性测度
MEASURES OF ASSOCIATION AND DEPENDENCE THROUGH COPULAS

```
86                              dbeta(b,v/by,(1-v)/by)) / length(u)
87         } }
88      return(data.matrix(mat)) }
89
90  library(copula)
91  #copula: copula model, must be a build-in function in R
92  #bx,by: bandwith of beta kernel
93  #n: sample size
94  rho.beta.esti<-function(copula,bx,by,n){
95  den <- copula.density(100,100,copula) # 100 grids in both x and y
         directions
96  set.seed(8192)
97  X <- rCopula(n,copula)            #generate random data from the
         COPULA defined above
98  grids <- 100                      #number of grids, beta estimator
         will be evaluated on those grids.
99  Z<- beta.kernel.copula.surface(X[,1],X[,2],bx,by,p=grids) # both beta
         kernel has bandwith 0.1
100 print(c("rho1    is",dens.rho1(den)))
101 print(c("rho1 estimator is",dens.rho1(Z)))
102 print(c("rho2    is",dens.rho2(den)))
103 print(c("rho2 estimator is",dens.rho2(Z)))
104 print(c("norm    is",dens.rho(den)))
105 print(c("norm estimator is",dens.rho(Z)))
106 u <- seq(1/grids, len=(grids-1), by=1/grids)
```

Appendix

```r
# the following code will draw a surface
# persp(u,u,Z,theta=30,border="black", col="green",shade=TRUE,box=
    FALSE,zlim=c(0,6))
# this code print a almost transparent surface
# par(mfcol=c(1,2))
# print(persp(u, u, den, theta = 30, phi = 30, expand = 0.5, scale =
    TRUE, border="black", col="white"))
print(persp(u, u, Z, xlab ="u", ylab ="v", main =c("beta estimation
    with N=",n),theta = 30, phi = 30, expand = 0.5, scale = TRUE,
    border="black", col="white"))
# par(mfcol=c(1,1))
# another printing
# persp(u, u, Z, theta = 30, phi = 30, expand = 0.5, col = NA, border
    ="green")
}

# draw 3D surface
# library(rgl)
# range <- c(0,1)
# persp3d(u,u,Z, col="green3", aspect="iso", xlim=range, ylim=range,
    zlim= 1.5*range, axes=TRUE, box=TRUE,
#           xlab="X-axis label", ylab="y-axix label", zlab="z-axix
    label",surface=FALSE)
# p = p0
# matrix1 = matrix(1,nrow = p-1, ncol = p-1)
# persp3d(u,u,matrix1, col="red", add=TRUE)
```

基于Copula的相关性测度
Measures of Association and Dependence Through Copulas

```
126
127 #compare with different sample size
128 compare.ss <- function(copula){
129     par(mfrow=c(2,2))
130     persp(copula,dCopula,xlab="u",ylab="v",main="copula density
            function", theta = 30, phi = 30, expand = 0.5, col = NA, border
            ="black")
131     rho.beta.esti(copula, 0.01,0.01,50)
132     rho.beta.esti(copula, 0.01,0.01,100)
133     rho.beta.esti(copula, 0.01,0.01,10000)
134     par(mfrow=c(1,1))
135 }
136     copula<-normalCopula(0.5)
137 # copula <- fgmCopula(1)
138 # copula <- gumbelCopula(2)
139 compare.ss(copula)
140
141 # add one persp to the previous persp
142 # copula<-normalCopula(0.5)
143 # u <- seq(1/grids, len=(grids-1), by=1/grids)
144 # set.seed(8192)
145 # X <- rCopula(1000,copula)          #generate random data from the
            COPULA defined above
146 # grids <- 100                       #number of grids, beta estimator
            will be evaluated on those grids.
147 # u <- seq(1/grids, len=(grids-1), by=1/grids)
```

Appendix

```
148 # Z<- beta.kernel.copula.surface(X[,1],X[,2],0.05,0.05,p=grids) #
       both beta kernel has bandwith 0.1
149 # persp(copula,dCopula,xlab="u",ylab="v",main="copula density
       function", theta = 30, phi = 30, expand = 0.5, col = NA, border="
       blue")
150 # par(new = TRUE)
151 # persp(u,u,Z,xlab="u",ylab="v",main="copula density function", theta
       = 30, phi = 30, expand = 0.5, col = NA, border="red")
152 #
```

rho_estimation_based_on_beta_kernel.R

D Kernel Estimation

```
1 library(ks)
2 set.seed(8192)
3 x <- rmvnorm.mixt(1000)
4 # y <- rmvnorm.mixt(100, mus=c(0,0), Sigmas=invvech(c(1,0.99,1)))
5 # z <- rmvnorm.mixt(100, mus=c(0,0), Sigmas=invvech(c(1,0.1,1)))
6 #
7 Denx <- kcopula.de(x)
8 # Denx$estimate has those values: Denx$estimate[0,89] numeric(0),
       Denx$estimate[1,89]
9 #   [1] 0.1518968, Denx$estimate[N,89] [1] 0.2059955, so it is 0 on
       lower and left border,
```

基于Copula的相关性测度
Measures of Association and Dependence Through Copulas

```
10  # and non-zero on higher and right border.
11  plot(Denx, disp="persp", thin=3, col="white", border=1)
12  #
13  # Chatx <- kcopula(x)
14  # plot(Chatx, disp="persp", thin=3, col="white", border=1)
15
16  #scatterplot of x and pseudo observations.
17  par(mfrow=c(1,2))
18  plot(Denx$x.orig)
19  plot(Denx$x)
20  par(mfrow=c(1,1))
21
22
23  # test whether the total sum of density is 1
24  # for(i in 1:N){
25  #   for(j in 1: N){
26  #     test <- test + Denx$estimate[i,j]*delta^2
27  #   }
28  # }
29  # print(test)
30  #[1] 0.9867988, it is not close to 1.
31
32  # estimate rho1
33  ################################################################
34  # Method one of estimating integral of copula density ( better)
35  # N <- nrow(Denx$estimate)
```

Appendix

```
36 # delta <- 0.006666667
37 # sum <-0
38 # rho1.sqr <- 0
39 # Denx$estimate[0,]<-0
40 # density.inte <- 0
41 # for(i in 1:N) {
42 #   for(j in 1:N){
43 #     density.inte <- 0
44 #     density.inte<-sum((Denx$estimate[i,1:j-1]+Denx$estimate[i,2:j
         ])*delta/2) + Denx$estimate[i,1]*delta/2
45 #     rho1.sqr <- rho1.sqr + density.inte^2*(delta)^2
46 #   }
47 # }
48 # rho1.sqr
49 #
50 # rho2.sqr <- 0
51 # for(i in 1:N) {
52 #   for(j in 1:N){
53 #     density.inte <- 0
54 #     density.inte<-sum((Denx$estimate[1:i-1,j]+Denx$estimate[2:i,j
         ])*delta/2) + Denx$estimate[1,j]*delta/2
55 #     rho2.sqr <- rho2.sqr + density.inte^2*(delta)^2
56 #   }
57 # }
58 # rho2.sqr
59 ########################################################
```

基于Copula的相关性测度
MEASURES OF ASSOCIATION AND DEPENDENCE THROUGH COPULAS

```
60  # Method two of estimating integral of copula density
61  # N <- nrow(Denx$estimate)
62  # delta <- 0.006666667
63  # sum <-0
64  # rho1.sqr <- 0
65  # Denx$estimate[0,]<-0
66  # density.inte <- 0
67  # for(i in 1:N) {
68  #   for(j in 1:N){
69  #     sum <-0
70  #     for(k in 1:j){
71  #       sum <- sum + Denx$estimate[i,k]*delta
72  #     }
73  #     rho1.sqr <- rho1.sqr + ((sum)^2)*(delta)^2
74  #   }
75  # }
76  # rho1.sqr
77  # # estimate rho2
78  # rho2.sqr <-0
79  # for(i in 1:N) {
80  #   for(j in 1:N){
81  #     sum <-0
82  #     for(k in 1:i){
83  #       sum <- sum + Denx$estimate[k,j]*delta
84  #     }
85  #     rho2.sqr <- rho2.sqr + ((sum)^2)*(delta)^2
```

Appendix

```
86  # }
87  # }
88  # rho2.sqr
89  ####################################################
90  # test whether the density estimation and distribution estimation are consistent.
91  # the result is they are not! and the difference is big.
92  # Denx$estimate[N,89]
93  # M <- 23
94  # (Chatx$estimate[M,M] + Chatx$estimate[M-1,M-1] - Chatx$estimate[M-1,M] - Chatx$estimate[M,M-1]) - Denx$estimate[M,M]
95  ##########################################################################
96  # function- calculate \rho 1^2 (Y is a function of X) from the estimated copula density function
97  dens.rho1 <- function(chat)
98  {
99    # chat= c(x,y) density
100   delta1 <- diff(chat$eval.points[[1]])[1]  # delta1 = delta x
101   delta2 <- diff(chat$eval.points[[2]])[1]  # delta2 = delta y
102   chat1 <- chat
103   rho1.sqr <-0
104   for(i in 1:nrow(chat1$estimate)) {
105     for(j in 1:ncol(chat1$estimate)){
106       density.inte <-sum((chat1$estimate[i,1:j-1]+chat1$estimate[i,2:j])*delta2/2) + chat1$estimate[i,1]*delta2/2
107       rho1.sqr <- rho1.sqr + density.inte^2*delta1*delta2
```

基于Copula的相关性测度
Measures of Association and Dependence Through Copulas

```
108        }
109      }
110      return(rho1.sqr)
111    }
112   # function- calculate \rho 2^2 (X is a function of Y) from the
          estimated copula density function
113   dens.rho2 <- function(chat)
114   {
115      # chat= c(x,y) density
116      delta1 <- diff(chat$eval.points[[1]])[1] # delta1 = delta x
117      delta2 <- diff(chat$eval.points[[2]])[1] # delta2 = delta y
118      chat2 <- chat
119      rho2.sqr <-0
120      for(i in 1:nrow(chat2$estimate)) {
121         for(j in 1:ncol(chat2$estimate)){
122            density.inte <-sum((chat2$estimate[1:i-1,j]+chat2$estimate[2:i
                 ,j])*delta1/2) + chat2$estimate[1,j]*delta1/2
123            rho2.sqr <- rho2.sqr + density.inte^2*delta1*delta2
124         }
125      }
126      return(rho2.sqr)
127   }
128   # calculate rho^2 (the norm of C) from the copula density estimator
129   dens.rho <- function(chat)
130   {
131      rho.sqr <- dens.rho1(chat) + dens.rho2(chat)
132      return(rho.sqr)
```

Appendix

```r
133 }
134
135
136 library(ks)
137 library(copula)
138
139 set.seed(8192)
140 x <- rmvnorm.mixt(100, mus=c(0,0), Sigmas=invvech(c(1,0.5,1)))
141 chatx <- kcopula.de(x) #copula density estimation
142 # par(mfrow=c(1,2))
143 persp(normalCopula(0.5),dCopula,theta = 30, phi = 30, expand = 0.5,
          col = NA, border="blue")
144 par(new = TRUE)
145 plot(chatx, disp="persp", theta = 30, phi = 30, expand = 0.5, col =
          NA, border="black")
146 par(new = TRUE)   #plot density estimator
147 par(mfrow=c(1,1))
148
149 persp(copula,dCopula,xlab="u",ylab="v",main="copula density function",
          theta = 30, phi = 30, expand = 0.5, col = NA, border="blue")
150 par(new = TRUE)
151 persp(u,u,Z,xlab="u",ylab="v",main="copula density function", theta =
          30, phi = 30, expand = 0.5, col = NA, border="red")
152
153 Chatx <- kcopula(x) #copula estimation
154 plot(Chatx, disp="persp", thin=3, col="white", border=1)
155 chatx <- kcopula.de(x) #copula density estimation
```

基于Copula的相关性测度
Measures of Association and Dependence Through Copulas

```
156 par(mfrow=c(1,2))
157 par()
158 Chatz <- kcopula(z)
159 plot(Chatz, disp="persp", thin=3, col="white", border=1)
160
161 plot(Chatx)
162 par(mfrow=c(1,2))
163 plot(chatx$x, sub="Pseudo observations")
164 plot(chatx$x.orig, sub="Original x") #chatx$x.orig is the same as x.
165 par(mfrow=c(1,2))
166 plot(x)
167 plot(chatx$x.orig, sub="Original x")
168 print(dens.rho1(chatx)) #calculate rho 1 from the density estimator
169 print(dens.rho2(chatx))
170 print(dens.rho(chatx)) #calculate norm of C from the density
        estimator, must run dens.rho1 and dens.rho2 first
171 # should be close to 2/3
172 Chaty <- kcopula(y)
173 chaty <- kcopula.de(y)
174 plot(Chaty)
175 plot(chaty)
176 print(dens.rho(chaty))  ## should be close to 1
177
178 library(copula)
179 theta <- 100
180 C2 <- onacopula("Clayton", C(theta, 1:2))
```

Appendix

```r
U <- rnacopula(n = 1000, C2)
plot(U, asp=1, main = "n = 1000 from Clayton(theta = 100)")
chatU <- kcopula.de(U)
plot(chatU, disp="persp", thin=3, col="white", border=1)

print(dens.rho(chatU))  ## should be close to 2/3

par(mfrow=c(3,2))
x <- rnorm.mixt(1000, mus=c(-1,1), sigmas=c(0.5, 0.5), props=c(1/2, 1/2))
y <- rnorm.mixt(1000, mus=c(-1,1), sigmas=c(0.5, 0.5))
z <- rnorm.mixt(1000, mus=c(-1,1), sigmas=c(0.5, 0.5), props=c(0.99, 0.01))
plot(x)
plot(y)
plot(z)
hist(x)
hist(y)
hist(z)
qua <- c(1/4,1/2,3/4)
mus <- c(-1,1)
sigmas <- c(0.5,0.5)

fhat <- dnorm.mixt(x=qua, 0, 1)
density <- pnorm(x,0,1)
density
```

Kernel_density_estimator.R

E FDM of variables in crime dataset

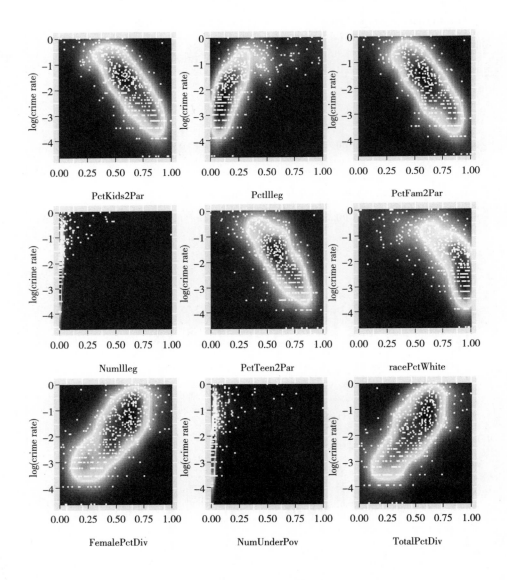